Introduction

How to Use This Unit

The information pages in this unit present science material at two levels of difficulty. Information for the teacher and/or older students is presented in a box on each page. Simplified information for younger students is presented under the illustration. You may wish to reproduce these pages to create a booklet.

About the Habitats in This Book

The animals mentioned in this unit often spend time in or near both the pond and the stream habitats.

No two ponds are exactly alike. The pond described in this unit demonstrates three characteristics common to numerous pond environments: plants ring its shoreline; its water temperature changes significantly each day during the warm season; and it is home to many living things.

It is not possible to include all the plants and animals that live in pond and stream habitats here. Make this unit more relevant for your students by including animals and plants that are specific to ponds and streams in your locality.

Preparation for Your Unit

1. Prepare an information center. (See page 2 for description)

2. Check your audiovisual catalog to see what films or videos are available for you to share with your class.

3. Explore community resources for possible speakers and plan field trip experiences.

4. Gather books from your school and public library to place in the information center. The bibliography on the inside cover of this unit contains some suggested resources, but is by no means a complete listing of what is available.

Room Environment

Information Center

Place a table in front of a bulletin board. Cover the board with butcher paper. Place a title on the board: *Let's Learn about Ponds and Streams.*

Use either side of the poster from this book as the focal point of the bulletin board. Leave space around the poster to display samples of student work from the unit. Put up pockets to hold activity sheets, writing paper, picture cards, etc., that the students can use.

Place books about pond and stream habitats and the animals that live in them on the table.

Bulletin Board

Create a stream with a strip of butcher paper attached to your classroom wall. Crumble the paper slightly before putting it up so that it has some dimension. Add paper rocks and bushes along the banks of the stream. In the places where the stream curves and at the end of the stream let the "water" spread out to form small ponds.

As your class learns about the members of the pond and stream communities, add to the stream and the pond. Your wall will become a pictorial record of everything that you have read and discussed.

Fill several large glass jars with pond water. Put the jars in direct sunlight and observe the changes in the water throughout your study.

Getting Started

What Is a Habitat?

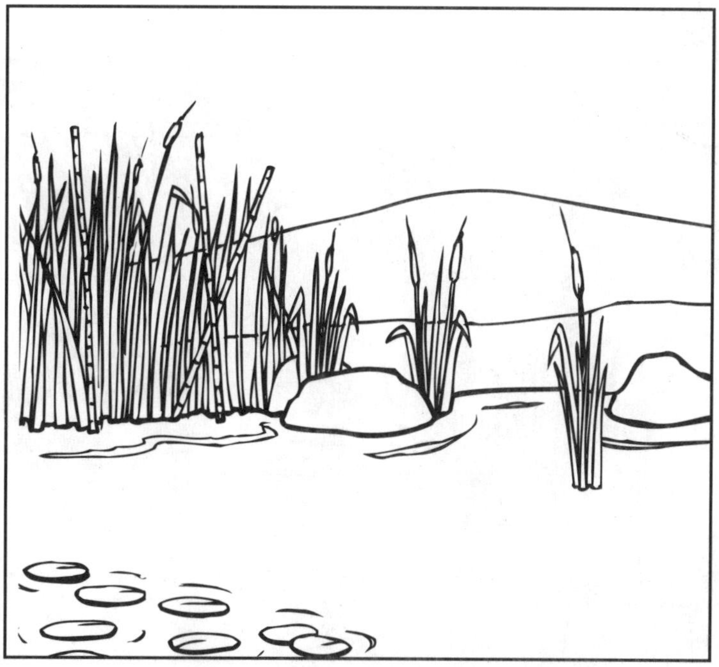

A habitat is a place where plants and animals naturally live. It is a place where there is air to breathe, good food to eat, water to drink, and shelter for protection.

The plants and animals in a given habitat have developed adaptations to enable them to survive in that habitat. The plants of the stream have a small compact shape and strong root connections so that the moving water cannot harm them, while many plants of the pond have large floating leaves so that they can absorb more sunlight.

An Introductory Discussion

What is a habitat?
How many habitats can you name?
Who or what lives there?
What does your habitat have that allows you to live and grow?
What makes a good habitat?
Do people live in only one kind of habitat?

Analyze Your Habitat.

Create a chart with four sections. Label the four sections: Air to Breathe, Food To Eat, Water to Drink, and Shelter for Protection. Ask students to tell how their habitat provides each of these essentials for life. Record their ideas on the chart.

Do a similar chart for a classroom pet or plant. Compare the results. How are the two habitats alike? How are they different?

Air to Breathe	Food to Eat
Water to Drink	Shelter for Protection

A Pond Habitat

What is a pond?

What plants and animals live there?

A pond is a shallow body of water with a muddy bottom. Aquatic plants usually grow from shore to shore. A pond might have been formed by the ancient work of glaciers moving across the land or a modern day landslide. It might have been formed as rivers and wind carved out a hollow spot. Or, maybe, the pond was created by animals or people.

Green plants growing around and in the pond capture energy from the sun and nutrients from the water. The plants provide food and a safe place for animals to live and grow. Small fish and insects hide among the leaves. They wait for prey or keep away from hungry predators.

Insects buzz and hum and creep and fly. Frogs and turtles sun along the bank and swim in the shallow water. Migrating ducks stop to rest and eat, and may even stay to raise a family. Other visitors come to drink, nibble the tender plants, and then return to their own habitats. Plants and animals depend on each other around the pond.

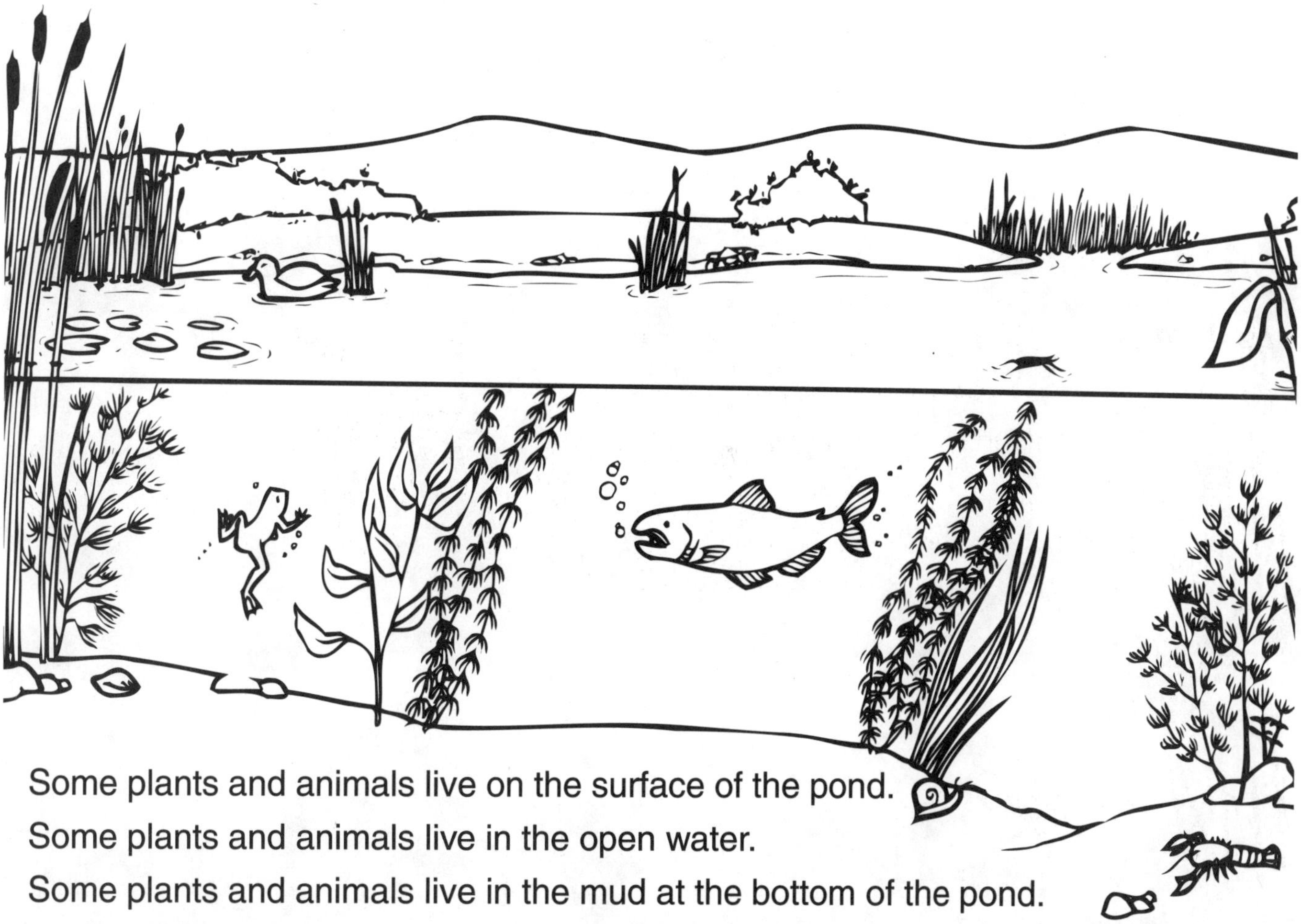

Some plants and animals live on the surface of the pond.
Some plants and animals live in the open water.
Some plants and animals live in the mud at the bottom of the pond.

Many forms of plants and animals live in layers in the pond. Some remain on the surface, some swim in the open water, and some live on the bottom.

In the ***muddy bottom*** of the pond, the decomposers are at work changing dead plant and animal remains into nutrients for new growth. The larvae of the whirligig beetle and the mayfly nymph graze on the bottom plants. Snails move slowly, scraping algae off rocks and plant stems. The tiny pill clam siphons algae and plant debris from the water as it rests on the bottom.

In the ***open water*** the sunfish and the bluegill dart among the underwater grasses. The backswimmer, a swimming insect, carries a bubble of air under its wings so that it can breathe under the water. Frogs, salamanders, and newts scoot through the water. They enjoy the warmth of the sun at the pond's edge where turtles are sunning themselves too. Beavers cut new trees at dusk to repair their dam.

Tiny animals make their homes on the ***surface film*** of the water. The water strider skitters along the surface and small snails hang from it. Mosquito larvae rest at the surface with their breathing tubes exposed to the air. Water spiders, whirligig beetles, and water treaders live and hunt on this fragile layer.

Note: Reproduce page 6 and 7. Have students cut out the examples of pond life and place them in the pond layer where they think that they live. Discuss with students why they chose to place plants and animals in the way that they did. Do not paste the cutouts down. Keep cutouts in an envelope as the class learns more about each pond inhabitant. At the end of the unit, paste the cutouts on the appropriate layers. Some of the plants are found at the water's edge and will not be placed in the water. At the end of the unit paste these on the back of the paper. (See answers on the inside back cover.)

Life in Layers

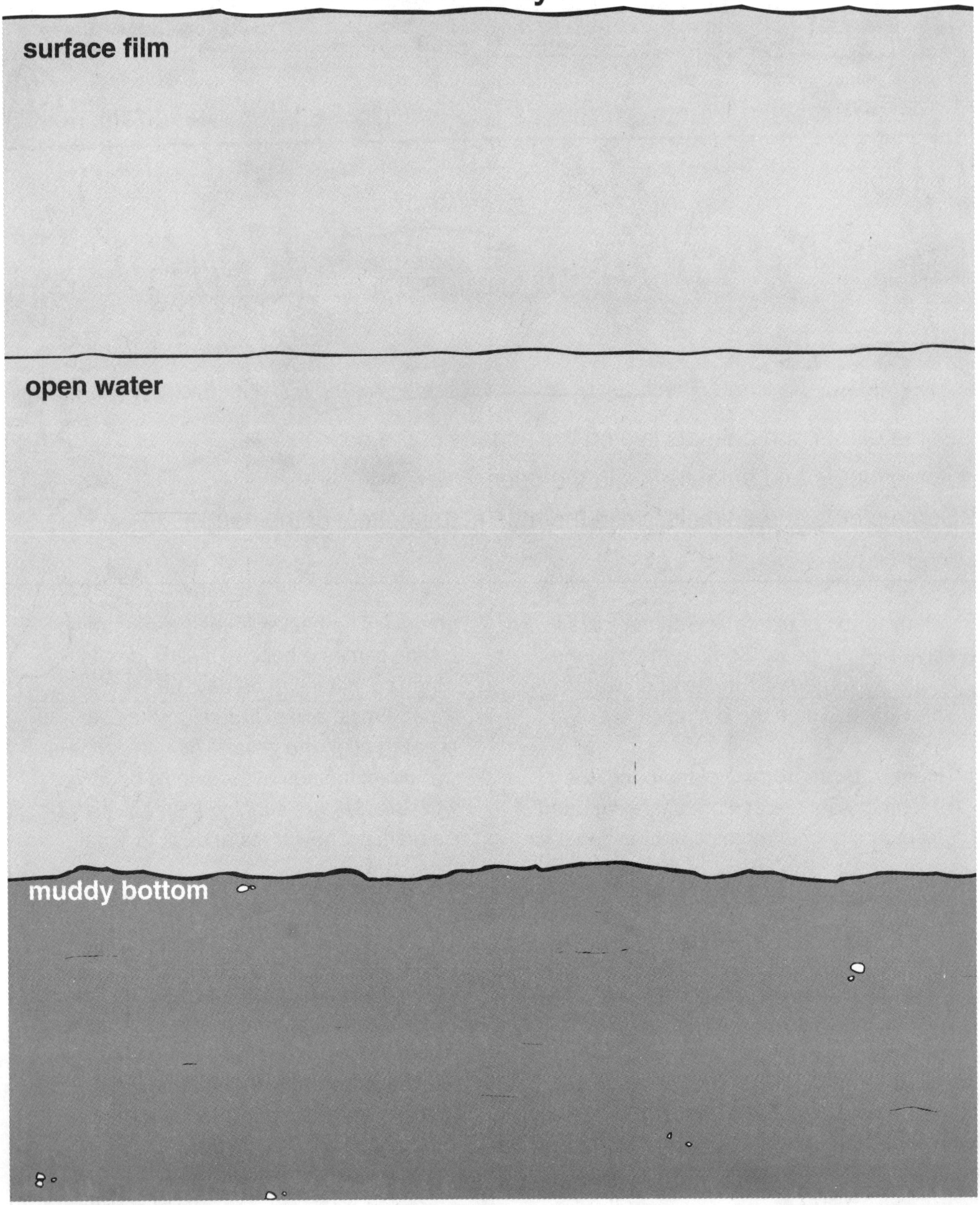

Note: to be cut and placed on layers of the pond.

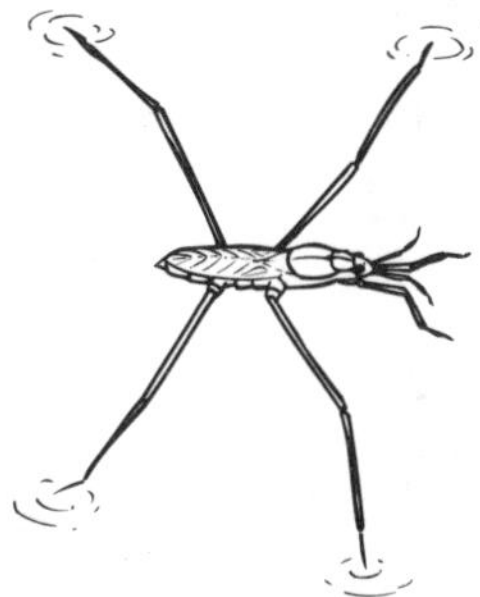
water strider

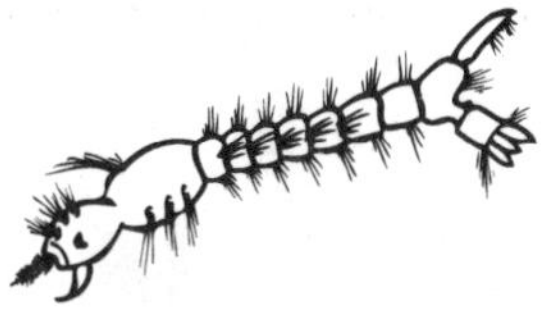
mosquito larvae

whirligig beetle

clam

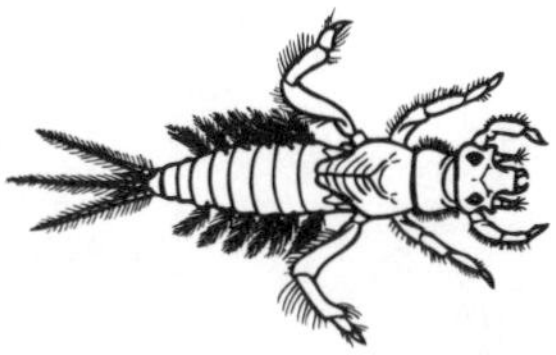
mayfly nymph

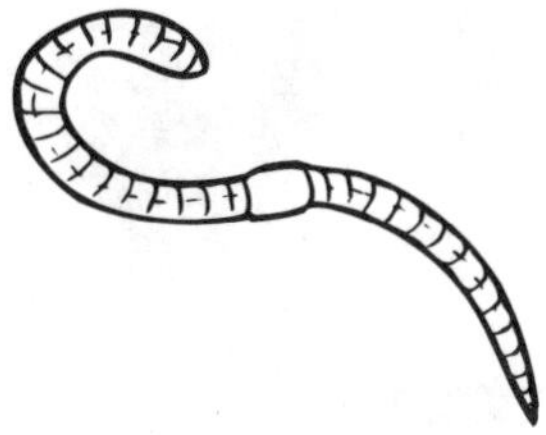
worm

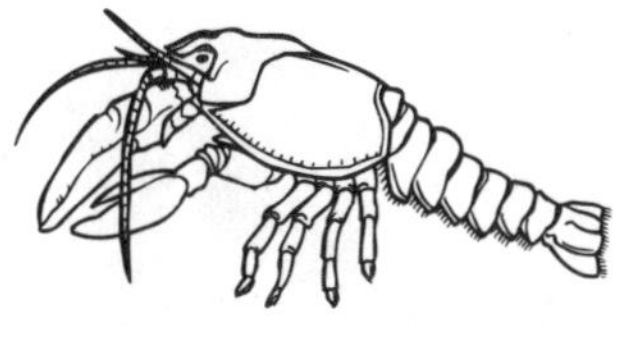
crayfish

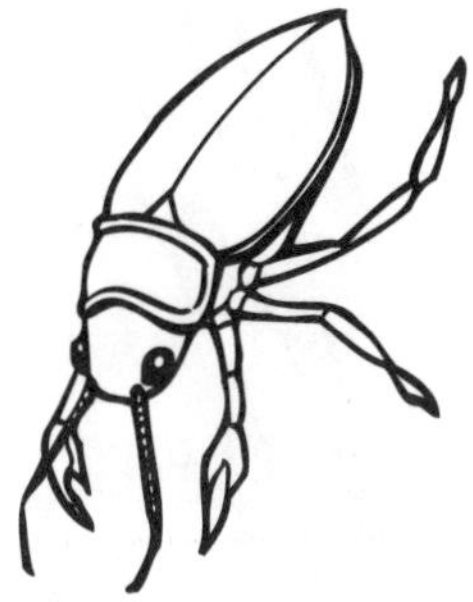
diving beetle

bass

sunfish

frog

beaver

duck

musk grass

bladderwort

water lily

duckweed

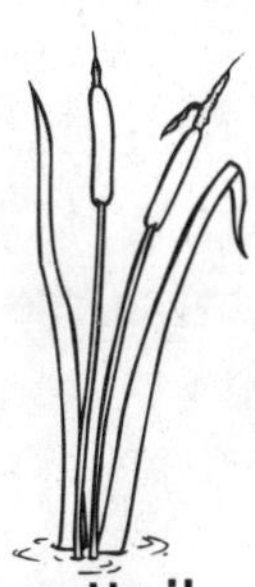
cattail

water fern

horsetail

Bottom Dwellers

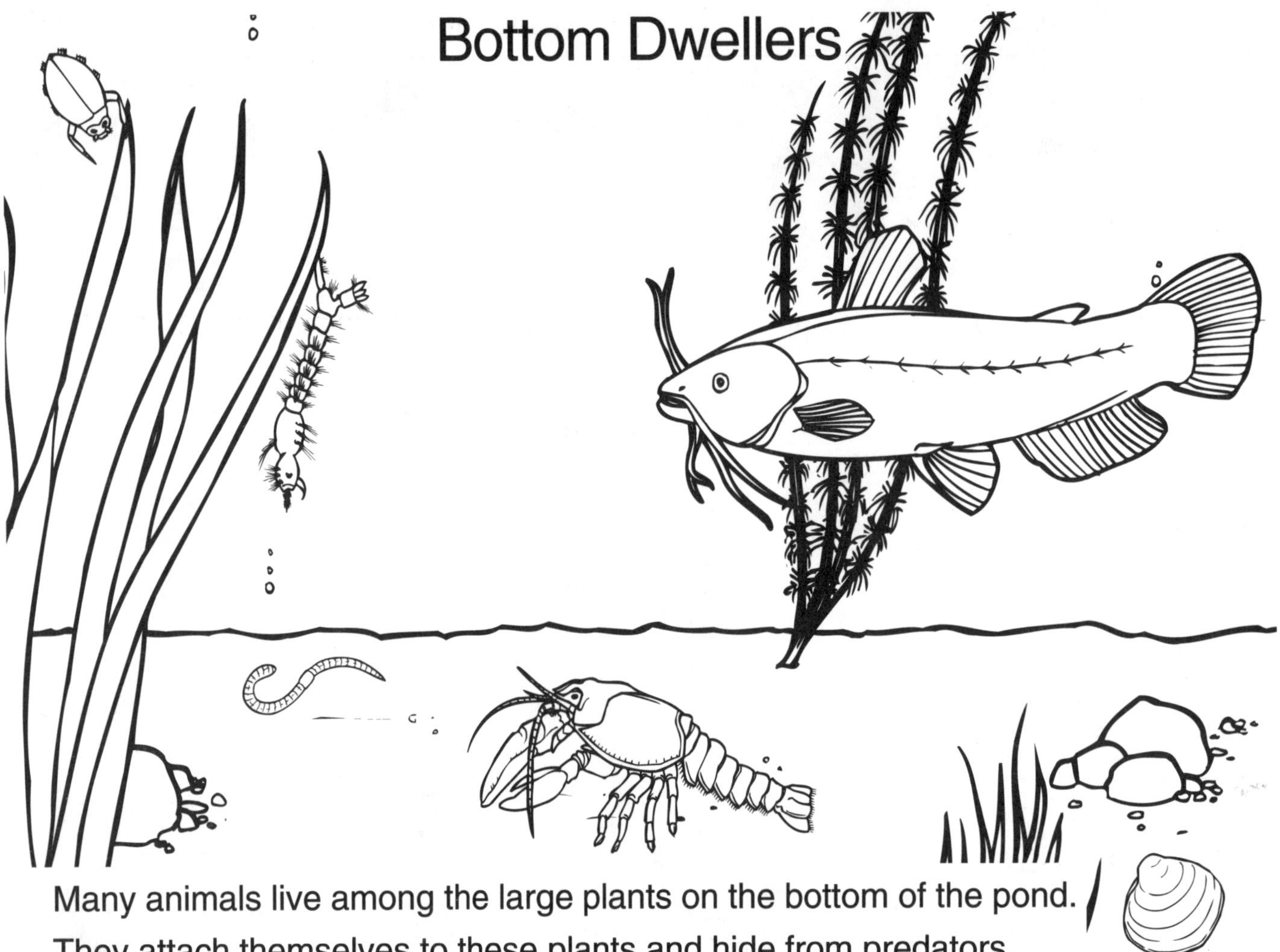

Many animals live among the large plants on the bottom of the pond. They attach themselves to these plants and hide from predators. Other animals live in the mud at the bottom of the pond.

Food at the bottom of a pond is abundant. ***Protozoans***, single-celled animals, are numerous. Tiny ***worms*** move through the mud, eating plants and animals. These animals are food for their larger predators, and their waste helps to recycle nutrients back to the plants. This is an important job in the pond community.

Tiny ***crayfish*** scurry on ten pairs of legs, using their two front claws for getting food. Larger animals and fish eat the crayfish, but they are hard to catch. Their stalked eyes can spot danger coming and they flee backward with a flick of their powerful tail.

The ***whirligig beetle larva*** crawls along the bottom plants. The ***caddis fly larva*** creeps by in its homemade house built of silk, saliva, and plant parts. It looks like a small pile of stems and leaves walking by.

The ***pill clam*** rests on the bottom. It siphons algae and plant debris from the water while it absorbs oxygen through its gills.

Catfish and ***carp*** scavenge the mud of the pond bottom for food. During the winter, turtles, frogs, and others hibernate in the bottom mud.

Note: You and your students will probably have wet feet if you are moving around a pond. Wear old shoes and have a dry pair ready for the return to school.

Visit a Pond Environment

Plan a trip to a pond to conduct the activities described below. Make the pond scope before the trip.

The basics of pond study are to simply keep alert and to sieve muck patiently. It is not necessary to kill and collect specimens in order to learn about them. They are better teachers if studied in their habitat or in an aquarium.

A Pond Scope

Make a pond scope to view the bottom. Take the bottoms off three empty juice cans. Fasten them into a long tube using strong, waterproof tape. Cover one end of the tube with heavy-weight transparent wrap. Secure tightly in place with rubber bands. Submerge the end with the transparent wrap and look down the inside of the tube. You can see clearly under the water as you probe the bottom.

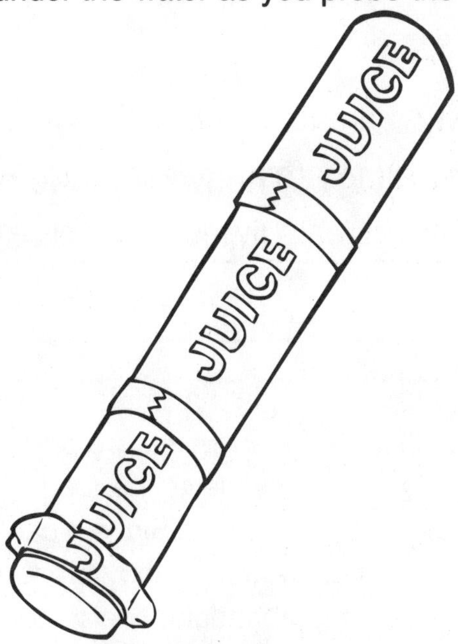

Map the Pond

While on site, mark the inlet and outlet of the pond. Use measurement skills as you record the size of the pond. Locate large rocks and logs. Note significant plant communities. List any animals that you observe. Note where you saw them.

If you can, visit the pond several times. Record the animal activity, changes in plant growth, and weather in a pond field journal. You may begin to see patterns in the growth and behavior of the pond life. Photographs and drawings are excellent for recording pond life during different seasons.

A Closer Look

Take an ordinary large metal food strainer. Scoop some mud and plant remains from the bottom of the pond. Place the mud in a white pan or bucket and begin gently separating the muck and plants from the animals. Have hand lenses available for viewing tiny organisms. Draw and list all creatures found. Be sure to return the sample to the pond after you have observed it.

Open Water

The water of the pond is filled with many animals.
Tiny animals filter fine food particles from the water.
These animals are the food for insects and fish.

Zooplankton, minute animals that swim and drift in the open water, are primarily filter feeders. They sieve fine food particles from the water. Since algae is a staple for these animals, they are found where algae is abundant. The ***water flea*** is a crustacean member of the zooplankton. It swims using its five pairs of legs. These legs also help to move water toward its mouth.

Many insects prowl the open water of the pond looking for a meal. Some of these insects seize their prey with their front legs and suck out juices. They have ingenious methods for getting air while below the water. The water scorpion thrusts its two hind legs above the surface and takes air in through special openings called ***spiracles***. Many pond insects can fly as well as swim.

The water of the pond is filled with fish. Many of the fish lay their eggs in the warm, shallow water at the edge of the pond. The ***sunfish*** and the ***large mouth bass*** stand guard until the eggs hatch, chasing off intruding fish and insects.

Larger animals eat the insects and the fish.
Some animals eat the plants that grow in the water.

Frogs, salamanders, and other amphibians dive through the open water of the pond. Tadpoles dart through the water and graze on the algae. Most tadpoles mature and transform into frogs within a year. Turtles, snakes, and alligators can also be found in the open waters of some ponds.

The abundance of water life and nesting sites attracts many birds to the pond. Shallow, weedy areas provide food for some ***ducks***, while others prefer to dive in the open water.

Mammals, too, frequent the pond. The ***beaver*** gnaws down trees and uses them to builds dams across streams and ponds. The new ponds created are a safe location for the beaver's home, called a lodge. The beaver propels itself through the water with webbed hind feet, using its scaly tail as a rudder. Other mammals like the ***muskrat*** and the ***raccoon*** rely on the pond as their food source.

The Water's Skin

The surface of water is held together by a strong force called surface tension. Surface tension makes water look as it if had a thin "skin" all over it. Help your students to see the surface film for themselves.

Viewing Surface Tension

- Fill a cup or a glass with water right to the brim.
- See how the surface tension pulls the surface of the water together so it curves above the rim of the container.

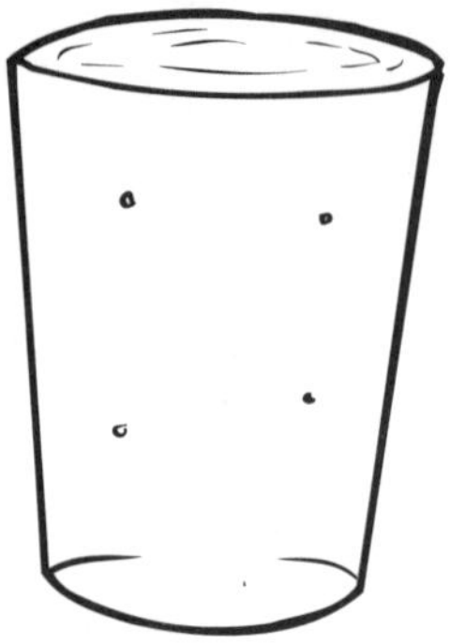

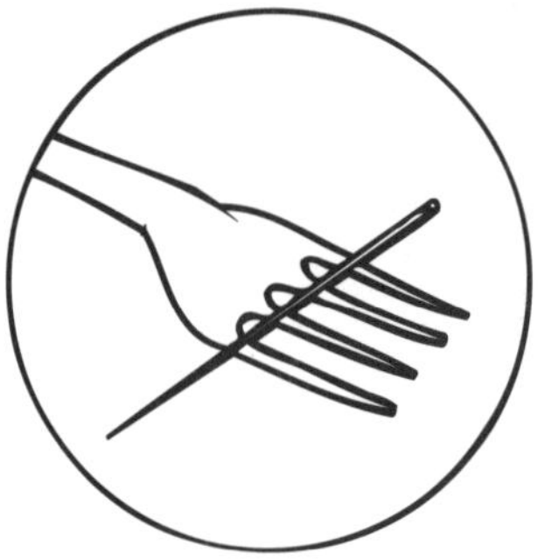

Float a Needle

- Fill a clean bowl with fresh water.
- Rest the needle across the prongs of a fork and gently break the surface of the water with the fork. If you are careful, the needle will float off as you take the fork away.
- Look closely—you can see the “skin” bending under the weight of the needle.

Insects that walk on the surface of ponds are supported by the water's "skin." The skin bends a little around their feet, just like it does around the needle; but it does not give way.

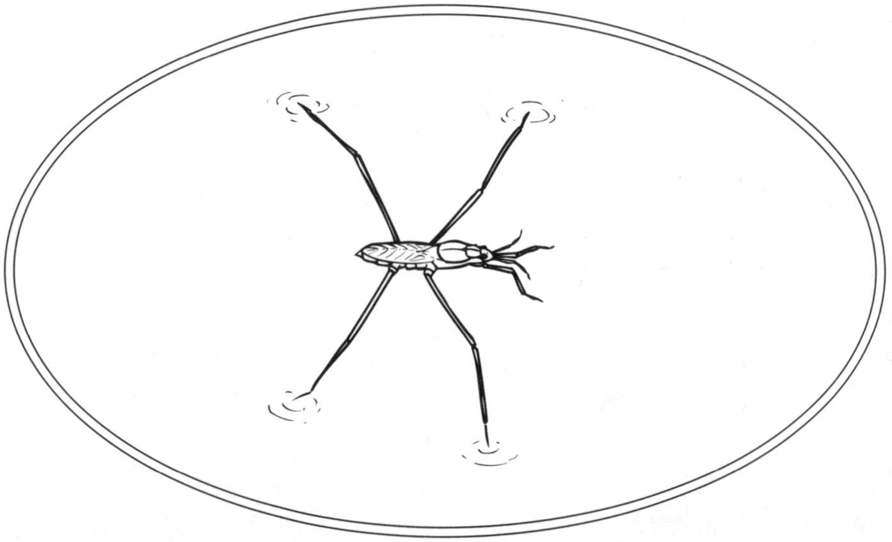

Surface Film

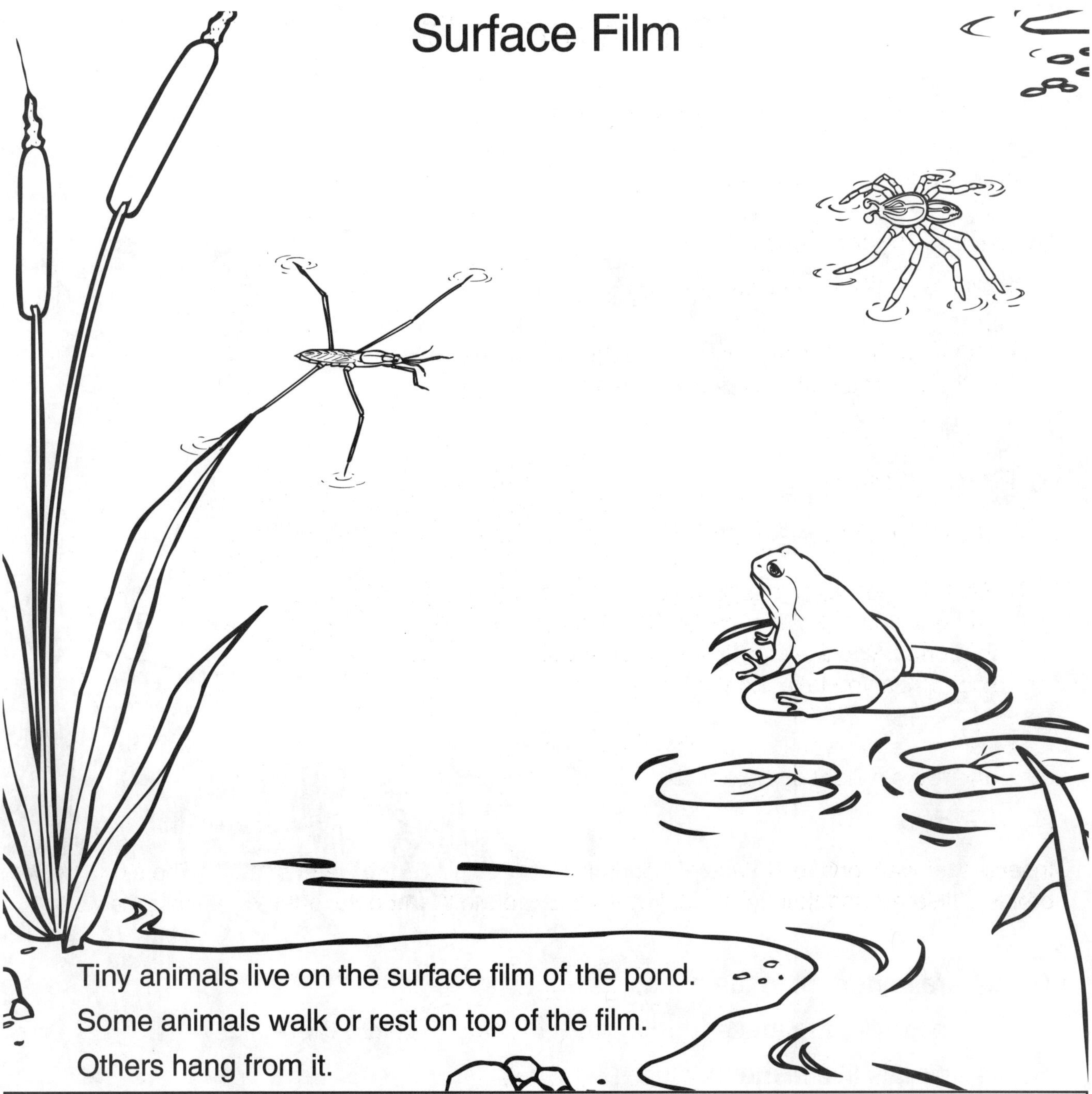

Tiny animals live on the surface film of the pond.
Some animals walk or rest on top of the film.
Others hang from it.

The surface tension of water molecules in the pond forms a delicate layer where tiny animals make their home. Scientists give the name ***neuston*** to organisms found on and under the surface film.

Mosquito larvae (wrigglers) rest at the surface with their breathing tubes exposed to the air. If disturbed, they quickly descend into deeper water. The wrigglers eat algae, microscopic animals, and debris. They have mouth brushes to filter food from the water.

The ***water strider, fish spiders,*** and ***water treaders*** walk on the surface of the water. ***Whirligig beetles*** use their legs as oars in their search for food. They have four sets of eyes—two that focus above the water and two that focus below the water.

Pond Plants

Plants are important to the pond.
They use sunlight to make food for the animals that live there.
Some animals live in the plants.

The pond world is dominated by plants. Through ***photosynthesis***, plants use sunlight to create growth that feeds pond life. Plants also provide homes for animals. The plant cover in and around the pond determines where the shade will be and how much of the water surface will be covered with leaves.

Plants also add oxygen to the water. Tiny green plants called ***algae*** are the most important plant life in the pond habitat. Some forms of algae float and drift in the water. Other forms are attached to rocks, stems, soil, and even some animals. Algae are the chief source of food in a pond.

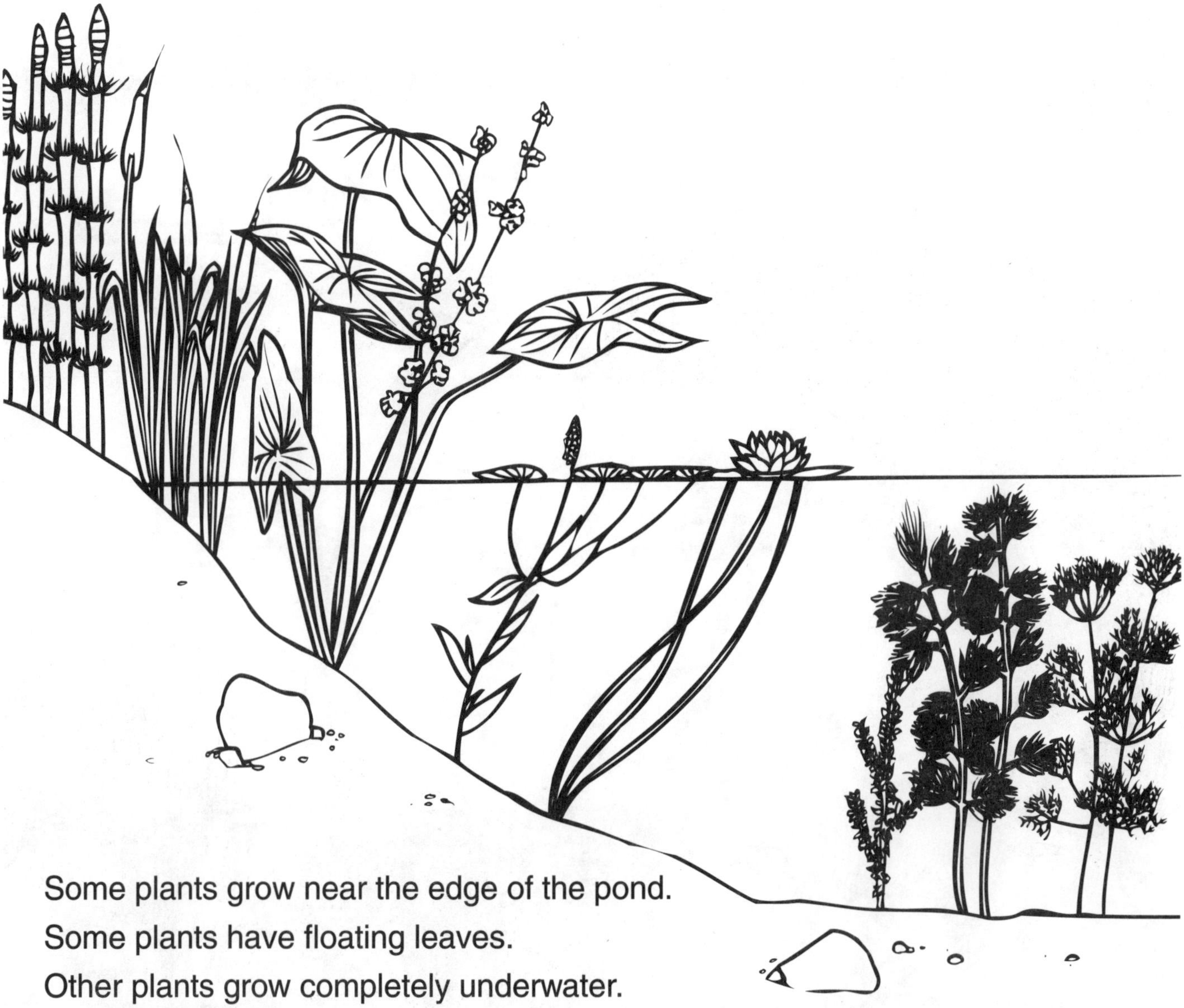

Some plants grow near the edge of the pond.
Some plants have floating leaves.
Other plants grow completely underwater.

Pond plants grow in different zones. The ***emergent zone*** is the area next to the edge of the pond. The horsetail and cattails, ferns, and sphagnum moss grow there. The plants are rooted on the bank or in the shallow water with their tops emerging into the air.

As the water deepens, the plants change. They are rooted underwater with long stems that allow their leaves to float on the surface of the water. This growing zone is called the ***floating-leaved zone***.

Water lilies, pond weed, water hyacinths, and wild celery reach for the surface there.

Duckweeds, tiny floating plants, do not have roots attached to the pond bottom. Their hanging roots receive nourishment directly from the water.

In deeper water the plants grow totally underwater. Bladderwort, elodea, and coontail are three common plants of the ***submergent zone***.

A Stream

The water in streams is always moving.

Living in a stream is different from living in a pond.

The water current makes it easier to do some things and harder to do other things.

Imagine living in the rushing water of a stream. The constant flowing water presents a challenge. Moving around can be difficult. Animals and plants can be dislodged and battered against rocks and logs. They might be covered with soil after a heavy rain, washed away during a flood, or even stranded on a dried-up stream bed.

When the water flows over a rough bottom, a stream becomes a giant mixer, saturating the water with dissolved oxygen. Flowing water is also filled with vital, dissolved nutrients. Most of these nutrients come from leaves that have fallen into the rushing water and soil that has eroded from the banks.

Plants that grow in the stream habitat have short stems and small leaves. They grow in a tight clump so the moving water can't wash them away.

Algae grows on the rocks, logs, and bottom of streams. It adapts to the stream habitat by growing in a flattened, tightly clumped mass with a sturdy, spreading attachment. Some stream algae produce new plants very quickly to replace those that are constantly being washed away by the current or eaten by the stream animals.

Larger plants of the stream bed tend to have short, flexible stems with small leaves. ***Watercress*** and ***liverwort*** are two examples.

Note: Create a chart with two columns, one headed ***Pond***, the other ***Stream***. Through discussion, decide where to place each card below. Ask – are there any similarities?

Differences in Plants

Still, shallow freshwater

Many kinds of plants. Some have floating leaves.

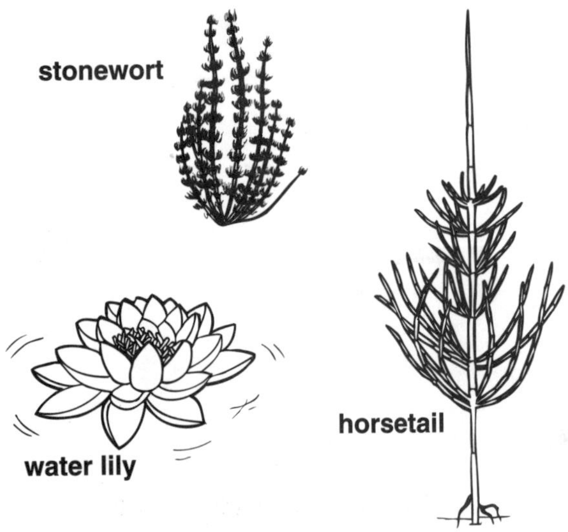

Flowing, freshwater

Plants have short, flexible stems. They grow in a tight clump.

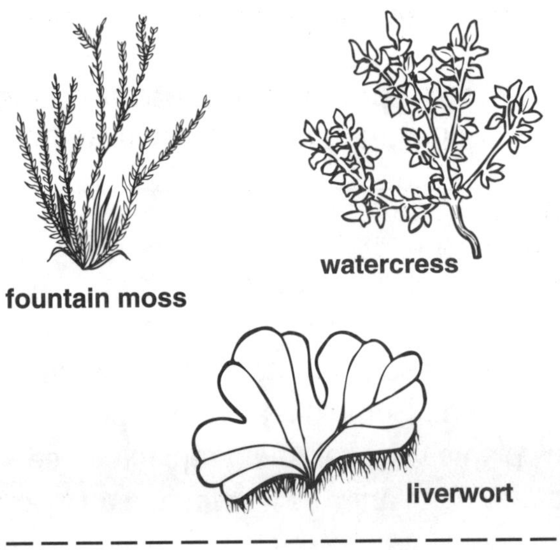

Collecting Pond Plant Specimens

While at the pond, collect plant specimens that can be pressed and used to create artwork.

Materials:

- pole with hook attached
- collecting bucket
- pond water
- flat pan or dish
- white paper
- newspaper
- cheesecloth
- heavy books

Steps to follow:

1. Fill a collecting bucket with pond water.

2. Using a hook on the end of a pole, reach under the surface and retrieve several submergents.

   Put the specimens in the water in the collecting bucket immediately.

3. Lay the drawing paper in a pan and cover it with pond water. Float and arrange one of the specimens over the paper.

4. Gently lift the paper from the water with the plant on top of it. Lay it on a newspaper. Cover it with cheesecloth, more newspapers, and finally several heavy books.

5. Let it dry overnight. The natural gelatins in the submergent will glue it to the paper. If the specimen is very heavy, you may have to apply some extra glue to its edges.

   Submergents, collected and pressed, make attractive cards, notepaper, placemats, and pictures.

The Frog

Leading a Double Life

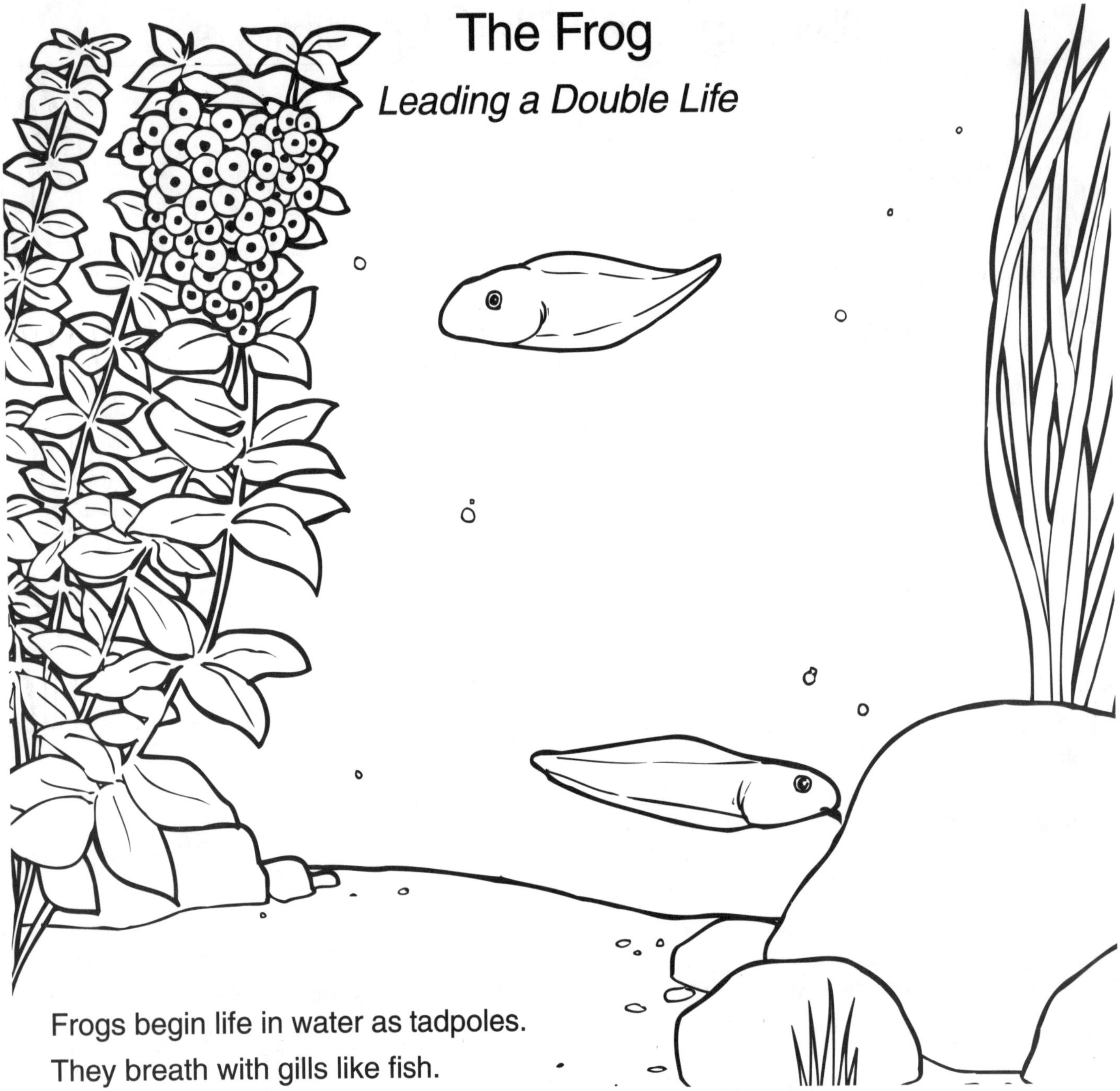

Frogs begin life in water as tadpoles.
They breath with gills like fish.
They eat algae.

As an ***amphibian***, a frog leads a double life. A ***tadpole*** hatches from an egg and starts swimming in the water. After going through some changes, it ends up a ***frog*** hopping on the edge of the pond.

In the spring, female frogs lay clumps of hundreds of jelly-covered eggs. When tadpoles hatch from the eggs, they swim and breathe through gills like fish. They get their oxygen from the water, not from the air.

They have long tails that move from side to side as they swim through the water hunting for food. Tadpoles eat algae. They scrape the algae off rocks and other plants with their tiny sharp teeth.

Tadpoles grow the back legs first.
Then they grow the front ones.
Soon they will look like frogs.

After a few weeks, legs will begin to grow—first, the two back ones; then, two front ones. The lungs start to develop so that tadpoles at this stage of development can live and breathe out of water. The tadpoles stop eating plants. Their tails are slowly absorbed by their bodies and serve as their nourishment.

At last the tadpoles become frogs. They are ready to leave the water and hop around on the land. But, because frogs have to keep their skin wet, they will stay near the pond.

A frog must keep its skin wet.
It stays close to water.
A frog's sticky tongue helps it catch an insect for its meal.

A frog doesn't drink water like we do. To keep from drying out, it jumps into a pond. The water goes right through its skin.

A frog's body temperature changes as the temperature of its surroundings changes. A frog warms up in the sun and cools off in the shade or in the water.

A hungry frog sits still and waits for his meal to come to him. His bulging eyes help him to see in all directions. His color helps him to blend into the background. When an insect moves, the frog flicks out his long, sticky tongue and captures his prey. Frogs also eat snails and worms.

Raising Frogs

If possible, before you start to study the life cycle of a frog, gather frog eggs from a pond. Observe what happens as they develop first into tadpoles, and then frogs.

1. Place the eggs in an aquarium open to the air. Fill the aquarium with water from the pond where they were found. Add pond plants.

2. Be sure that the aquarium is in direct sunlight so that the green algae will grow, providing food for the tadpoles when they hatch.

3. Observe the tadpoles after they hatch from the eggs. Notice how they change.

4. When they begin to develop legs, the tadpoles need a place where they can crawl out of the water. A rock or a sandy beach will keep them from drowning.

5. When the frogs are ready to leave the water, return them to the pond where the eggs were collected. They will be happier in their natural habitat.

The Life Cycle of a Frog

Note: Reproduce this page and the next page for each student. Students fill in the missing words. Cut both pages apart and match each picture with its caption. To display, lay pictures in order from egg to frog as shown. Flip pictures over and place glue on backs. Lay a piece of string or yarn down the center of the pictures, leaving enough at the top to tie a loop. Lay the correct caption on top of each picture so that the string/yarn is sandwiched between. Hang your frog life cycles.

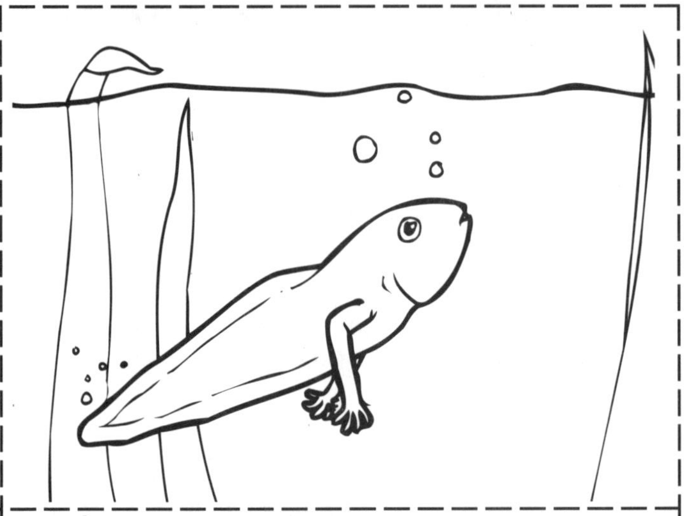

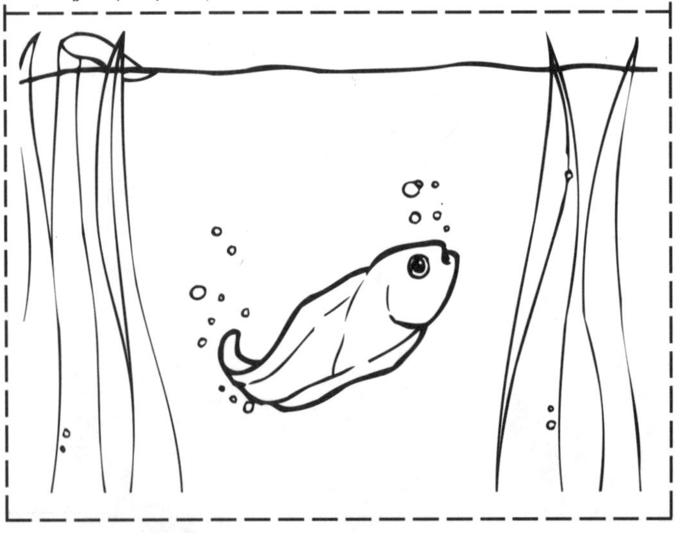

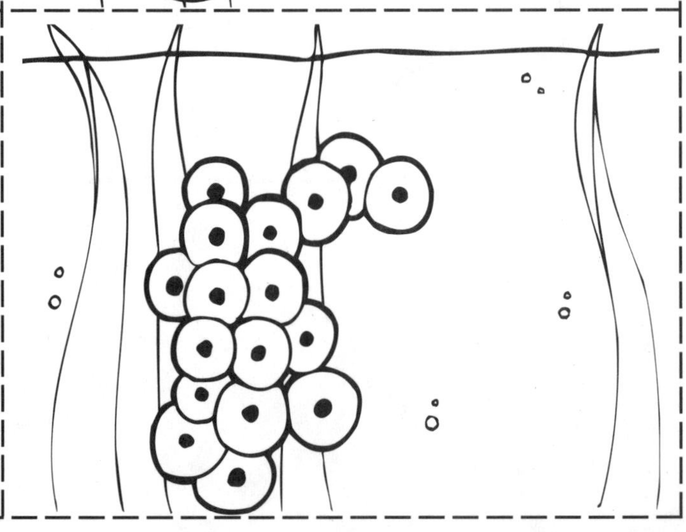

sunfish
stonewort
open water
muddy bottom
crayfish

horsetail
stonewort

oving Water

ants like
vater

watercress

n

liverwort

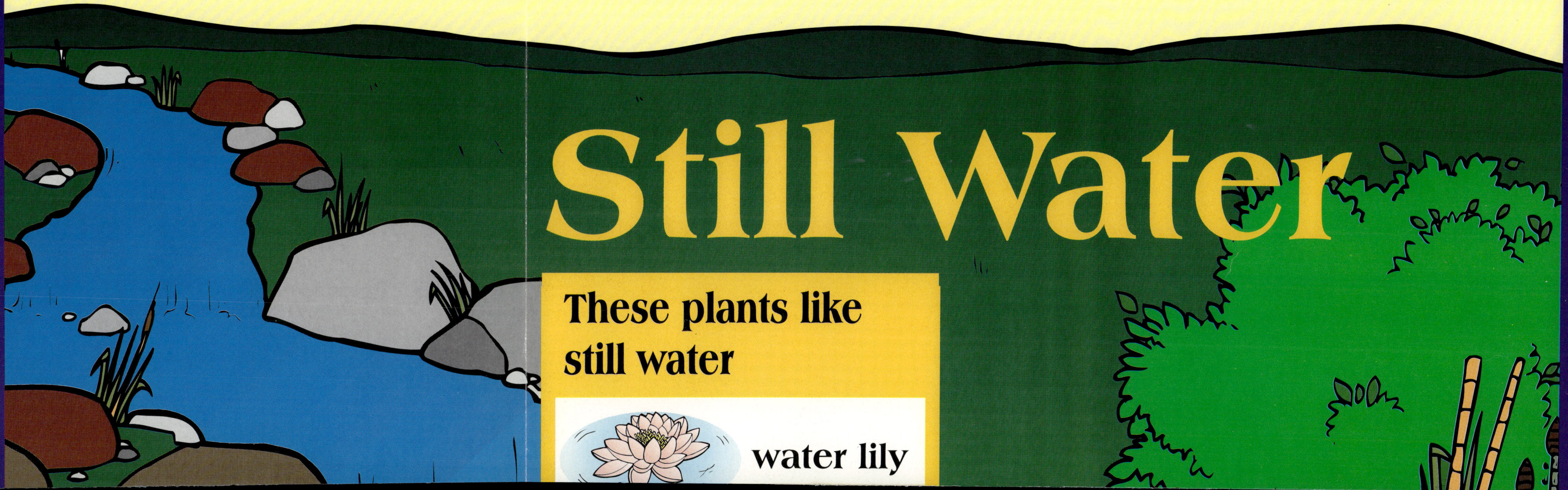

Still Water

These plants like still water

water lily

A Pond

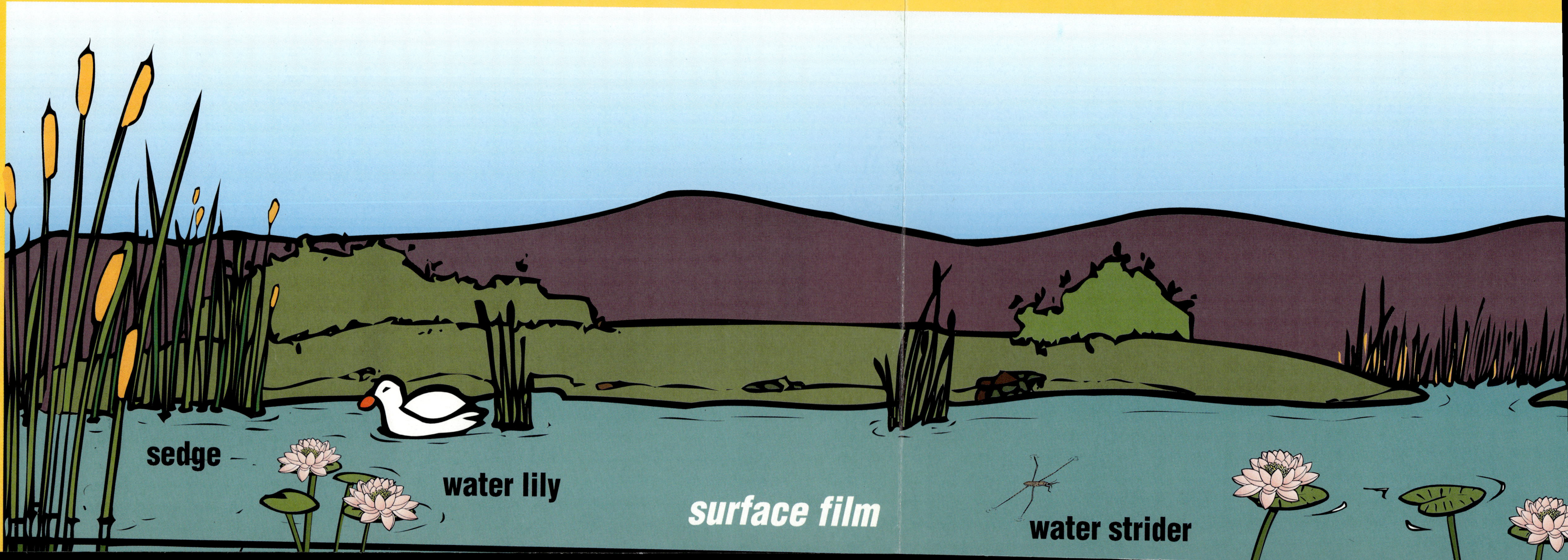

At last the tadpole has become a frog. It can live on land, but still stays close to water.	Now the tadpole has __________ legs. It is beginning to look like a __________________.
Tadpoles look a lot like fish. They breath with ________ like fish.	The tadpole has lungs and can breathe out of the __________________. Its ________________ gets shorter and shorter.
After a few weeks, legs begin to grow. The ________ legs grow first.	Jelly-covered frog eggs will soon hatch into _________.

Writing About Frogs

Write a diamonte poem to compare tadpoles and frogs. A diamonte poem usually shows change. The beginning line and the last line are opposites or contrasting words. The poem shows gradual change from the first line to the last line as it forms the shape of a diamond.

Reproduce the writing form on page 27 for each child. Work on one line at a time. Brainstorm and list words that describe tadpoles and frogs. For individual diamontes, students pick the words they like and write them on their forms. Or write one class diamonte and have students copy it onto their forms.

Read the diamonte aloud. Choose different groups of students to read different lines and create a choral reading.

Line 1 — one noun — *Tadpole*

Line 2 — two adjectives that describe a tadpole

Line 3 — Three participles (words ending in -ing) telling how a tadpole moves or what it does

Line 4 — Four nouns — the first two tell about tadpole and the second two tell about a frog

Line 5 — Three participles telling how a frog moves or what it does

Line 6 — Two adjectives that describe a frog

Line 7 — One noun — *Frog*

Tadpole
newborn, tiny
eating, growing, changing
swimmer, gills, hopper, lungs
croaking, sunning, leaping
spotted, bug-eyed
Frog

Diamonte Writing Form

Turtles

Reptiles with a Shell

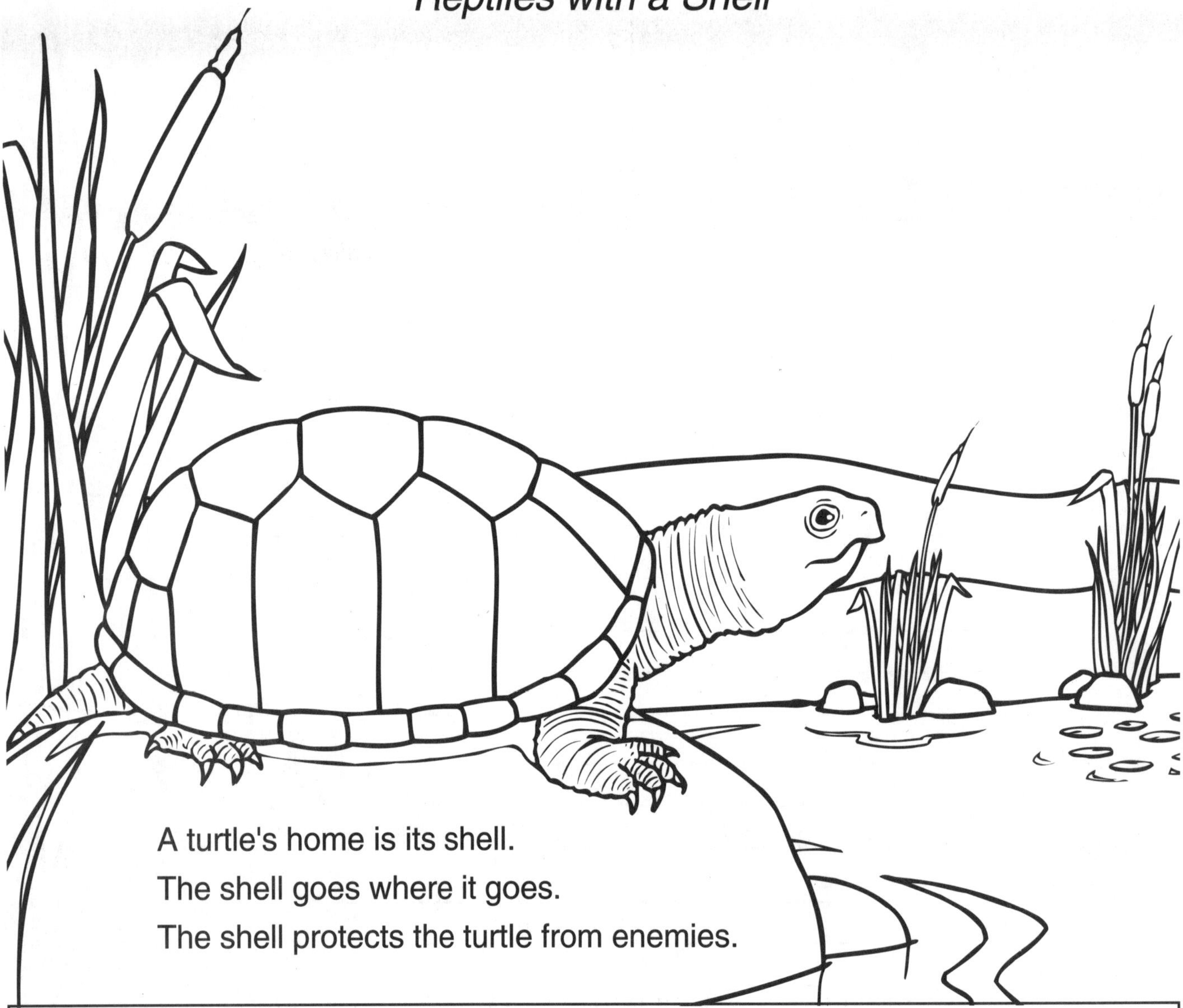

A turtle's home is its shell.
The shell goes where it goes.
The shell protects the turtle from enemies.

Turtles have been around for a long time. They may have been crawling about on earth even before their relatives the dinosaurs were born. Unlike other members of the reptile family, turtles have two shells joined at the sides. The upper shell is attached to its backbone and ribs. A turtle always carries his home with him and never leaves it.

With their short legs and wide shells, turtles move slowly on land. Those who swim, move quickly and easily in the water. Turtles breath with lungs like other reptiles.

The plates on a turtle's shell grow a new layer each year. The outside layers slowly wear off.

Turtles commonly found in ponds include the ***eastern painted turtle***, the ***spotted turtle***, and the ***mud turtle***.

Mother turtles lay their eggs in a hole that they dig.
Baby turtles look like small adults—shell and all.

Turtles are ***cold blooded*** like other ***reptiles***. You might say that they "run on solar energy." When their bodies are warmed by the sun they are able to move more easily. Around the pond when you see turtles resting in the sun on rocks or logs, they are "getting their batteries recharged."

Turtles have no teeth, but their jaws are bony and strong. They eat insects, worms, fruit, and plants.

Mother turtles dig holes and lay their eggs in the ground. They cover them to keep them warm and safe. When the turtles hatch they look like their parents, only smaller.

Turtles in Fact and Fiction

Read several books about turtles. Include nonfiction books and fiction books.

As a class, make a list of facts that you know about turtles. Write each fact on a section of a giant turtle shell.

Students can make and decorate flaps to cover the facts.

The result is a giant interactive turtle bulletin board.

Materials:

- green butcher paper
- white and yellow 9" x 12" construction paper (23 x 30.5 cm)
- stapler
- black marker

Steps to follow:

1. Cut a large turtle from green butcher paper, using simple shapes.
2. Round corners of white and yellow paper.
3. Write turtle facts on white paper using marking pen.
4. Staple as many facts as will fit on shell.
5. Staple yellow construction paper flaps over facts.

Make a Creeping Turtle

Materials:

- one "cup" of a paper egg carton
- yarn or string — about one yard long
- turtle body pattern
- brown or green construction paper—3" x 4"(7.5 x 10 cm)
- paints or colored markers
- black marker
- glue

Steps to follow:

1. Cut one cup off an egg carton.
2. Trim the bottom edge so that it is even all the way around. This will be the turtle's shell.
3. Paint the shell with tempera paints or color it with markers.
4. Poke a small hole through the top of the shell. Make a knot on the end of a piece of yarn. Push the yarn through the hole from the inside so that the knot is inside the shell and the long part of the yarn is coming out of the top of the shell. You will "walk" your turtle using this yarn.
5. Trace the shape of the turtle body on a piece of brown or green construction paper. Cut out the turtle's body. Fold the head up and draw eyes and a mouth.
6. Glue the turtle shell to the body. Let it dry.
7. Take your turtle for a walk.

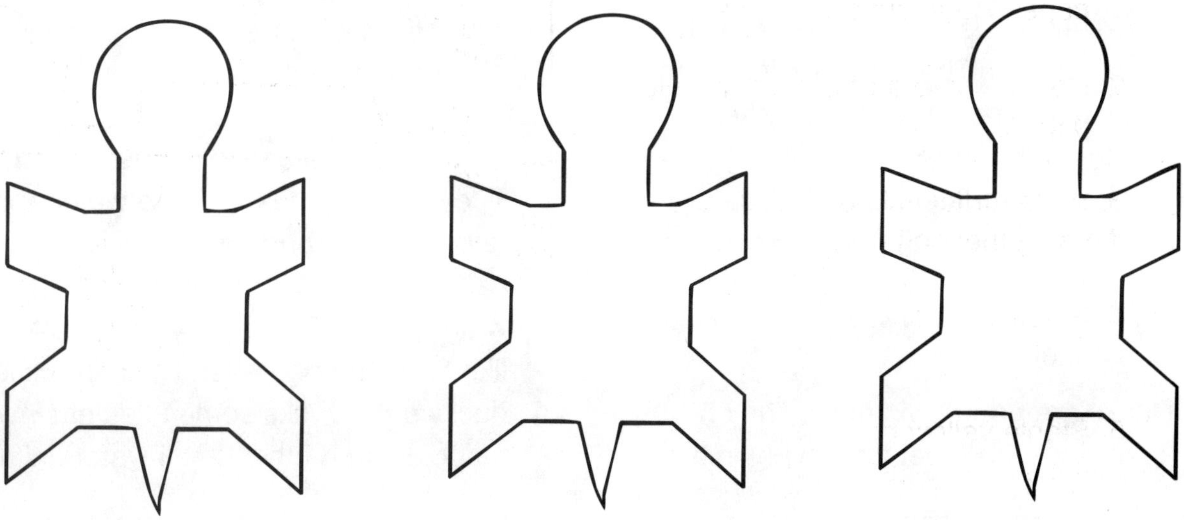

Ducks

Dabblers and Paddlers

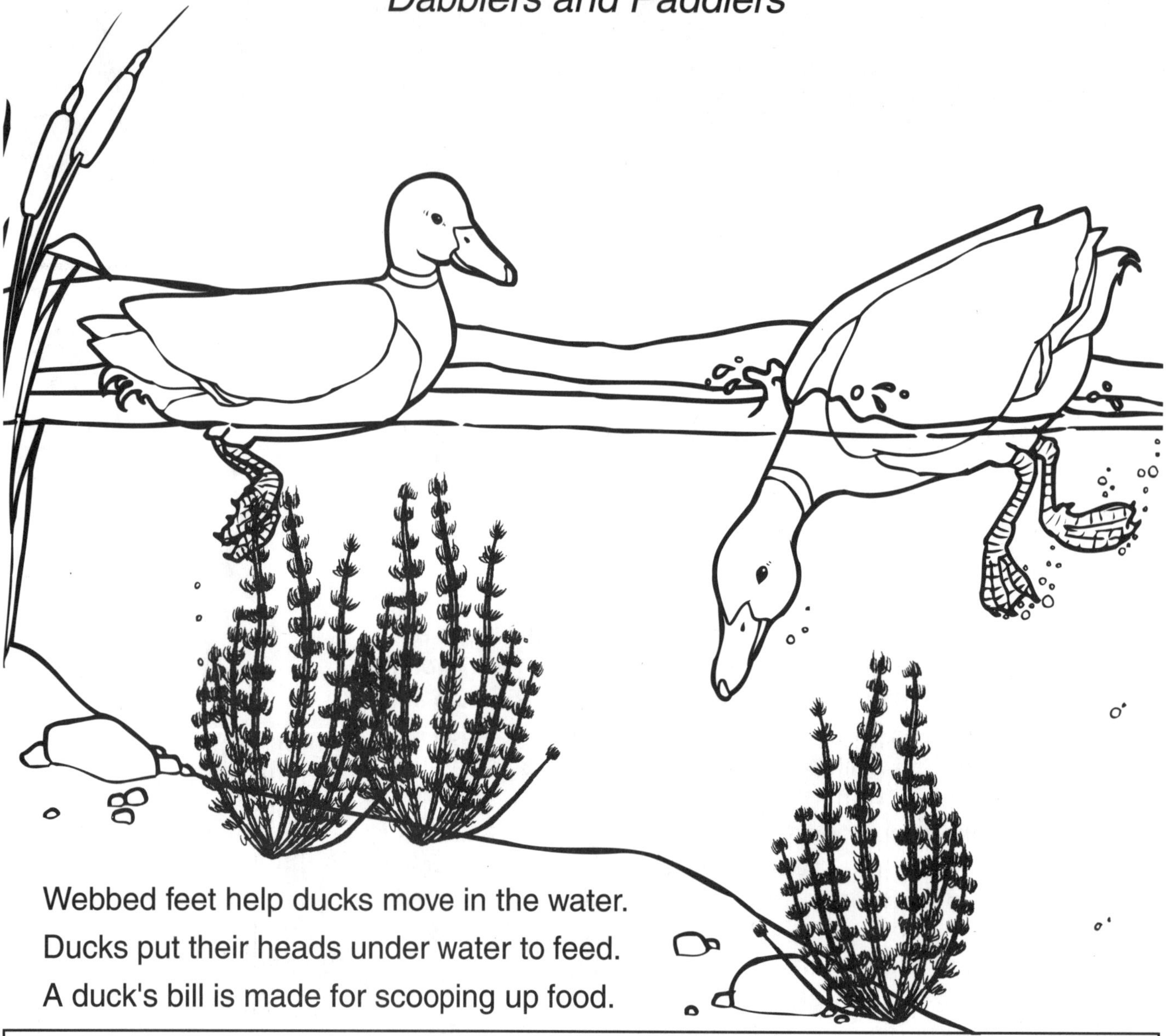

Webbed feet help ducks move in the water.
Ducks put their heads under water to feed.
A duck's bill is made for scooping up food.

Ducks are birds and, like all birds, they have feathers. They spend a lot of time in the water. When they smooth their feathers (preen) with their bills, they are making them waterproof. Ducks have an oil gland by their tail. They spread the oil over their feathers to keep the water away from their bodies.

Ducks are good swimmers. They have special webbed feet for moving in the water. Some ducks are divers. They can dive deep to get food.

The ducks you see in ponds are dabblers. They put their heads in the water and their tails in the air and paddle as fast as they can. They are feeding on weeds and water plants. There is even a plant they like to eat named after them. It's called duckweed. Ducks sometimes eat insects, tadpoles, and other small water animals.

Ducklings follow their mother to the pond.
She will help them learn to swim and find food.
In two months the ducklings will be on their own.

The ducks you see at the pond may be mallards. The male has a shiny dark green head. The female is brown. Her coloring helps her to blend into the surroundings. This is important when she is sitting on her nest incubating eggs.

When it is time to lay eggs, the female duck finds a place near the water. She pulls nearby grasses together to make a nest. She uses soft feathers from her breast to line it. Then she lays eight to ten eggs.

When the ducklings hatch, they are covered with a soft fuzz. As soon as their feathers dry, they begin walking, swimming, and looking for food. In two months, when their feathers have grown, they are ready to take off on their own.

As the days get cold and the food supply is short, ducks migrate south where it is warmer. They can fly for long distances on their strong wings.

Duck Dialogue

Preparation:

Read several books about ducks to your class.
Be sure to include both nonfiction and fiction selections.

Lead the class in a discussion of what ducklings might be saying to each other as they swim about in the pond. Direct the discussion toward factual information the class has learned about ducks.

Depending on the ability levels of your students, you may want to write words, phrases, or entire sentences on the board.

Materials:

- white paper — 12" x 18" (30.5 x 46 cm)
- blue paper for pond — 9" x 12" (23 x 30.5 cm)
- copy of page 35 with ducks and speech bubbles
- scissors
- glue
- colored paper scraps
- crayons

Steps to follow:

1. Guide students to plan what will go in the speech bubbles before beginning to write. Write in the bubbles before cutting them out.
2. Color and cut the ducklings.
3. Create a pond by pasting a free-form shape made from blue construction paper to a sheet of white paper.
4. Paste the ducks on the pond so they look like they're swimming.
5. Paste speech bubbles near each duckling.
6. Add details to the pond with colored paper scraps and crayons.

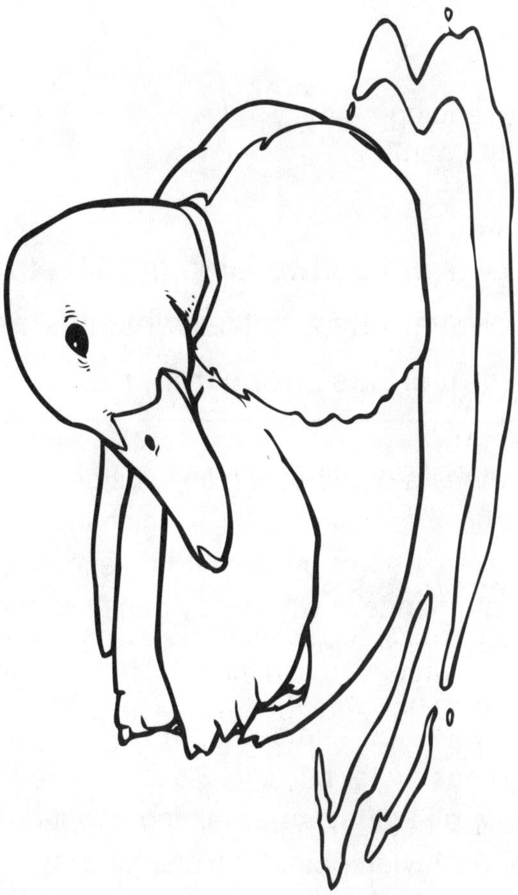

Beavers

Engineers of the Pond

A beaver has two large front teeth for chewing.

A beaver has a flat tail which helps it balance and swim.

A beaver has strong front paws for digging and carrying.

Beavers are interesting animals because they change the habitat in which they live. Beavers do this by blocking up streams to create ponds in which they can build their homes, called lodges.

Beavers' bodies make them well-suited for underwater building. Special muscles close off their nose, ears, and throats to keep the water out. Beavers can stay underwater for fifteen minutes while they are working. They have see-through eyelids that act like goggles so they can see well as they swim beneath the surface.

Beavers' broad tails act like rudders for steering. Their webbed back feet make them excellent swimmers. Strong front paws with claws help them dig and carry. Their two very large, orange front teeth are used to gnaw down trees.

Beavers cut down trees to build their dams.

A beaver dam is made with branches, stones, and mud.

The dam will make a pond on the stream.

When a beaver builds a dam, everyone in the family helps. Mother, Father, and three or four younger beavers work together. First they cut trees down with their big strong teeth. Next they gnaw them into smaller sections and drag them into the water.

They begin building their dam from the side of the stream, moving out toward the center. Stones are rolled and pushed on top of the limbs and branches to keep the wood from floating away. Beavers use their front paws to scoop mud from the bottom to fill in the spaces.

When the dam is completed, a pond will form behind it. The beavers will continue to check for leaks and repair the dam as needed. They are now ready to start work on the dome-shaped lodge that will be their home in the pond. The lodge looks like a giant pile of sticks from the outside. It is hollow inside, with a ledge above the water level. The entrance is under the water.

Note: Reproduce pages 38 and 39 for each child.

A Beaver Builds a Dam

A Pop-up Book

Materials:

- a copy of the dam-building process below
- 4 copies of the pop-up form on page 39
- scissors, paste, and crayons

Steps to follow:

1. Make 4 pop-ups.
2. Color and cut out the pictures. Paste one picture piece to each pop-up form. Paste the matching text on each page.
3. Paste the pop-up pages back-to-back as shown on page 39. Be sure they are in the correct order.
4. Fold a 6" x 9" (15 x 23 cm) construction paper sheet in half for a book cover. Paste the pop-up pages inside.

A beaver cuts tree limbs and branches.

The branches are drug to the site of the dam.

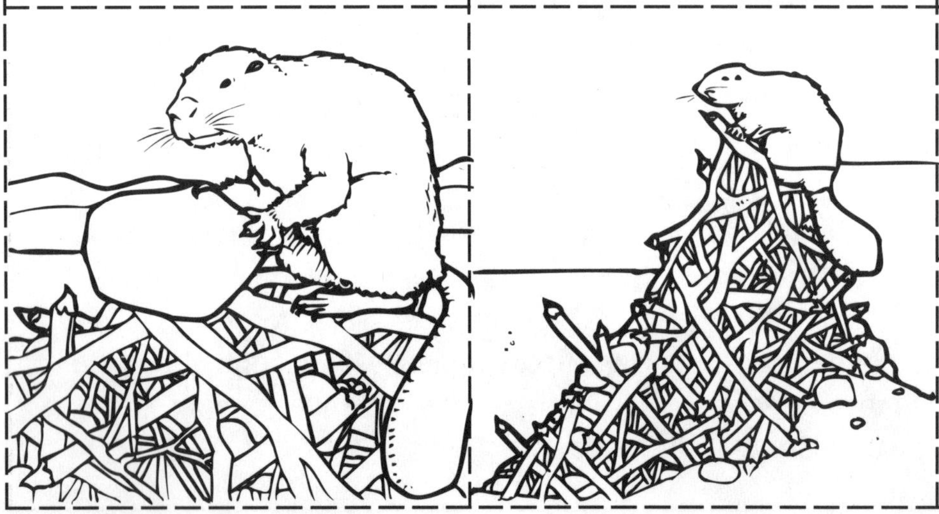

Stones are pushed on top of the wood.

The dam is finished and the beaver family can begin their lodge.

Note: An adult will need to paste the pages back-to-back for younger children.

Basic Pop-up Form

Each child will need
4 copies of this pattern.

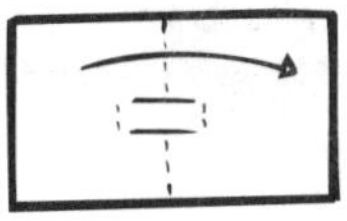

1. Cut out the basic form. Fold on 1.

2. Cut on dotted lines.
3. Fold on 2.

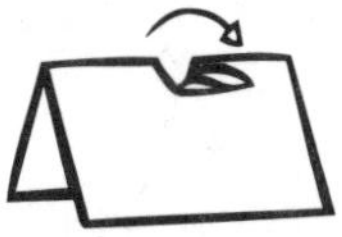

4. Open the paper and push the tab to the reverse side thereby reversing fold 2.

5. Pinch and pull the pop-up tab from the inside. Close and press the folds.

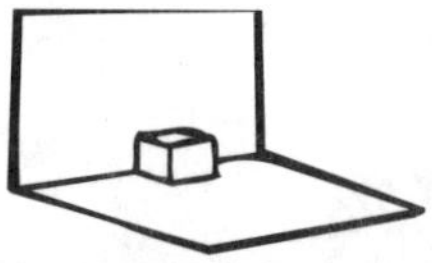

Make a Pop-up Book

Each child will need a copy of page 38.

1. Paste one picture and accompanying sentence to each pop-up form as explained on page 38.

2. Paste the pop-up pages back-to-back as shown. Be sure they are in the correct order.

3. Fold 6" x 9" (15 x 23 cm) construction paper in half for a book cover. Paste pop-up pages inside.

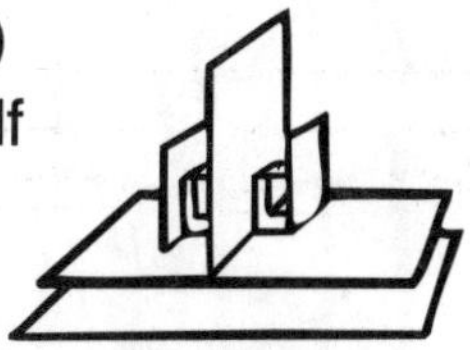

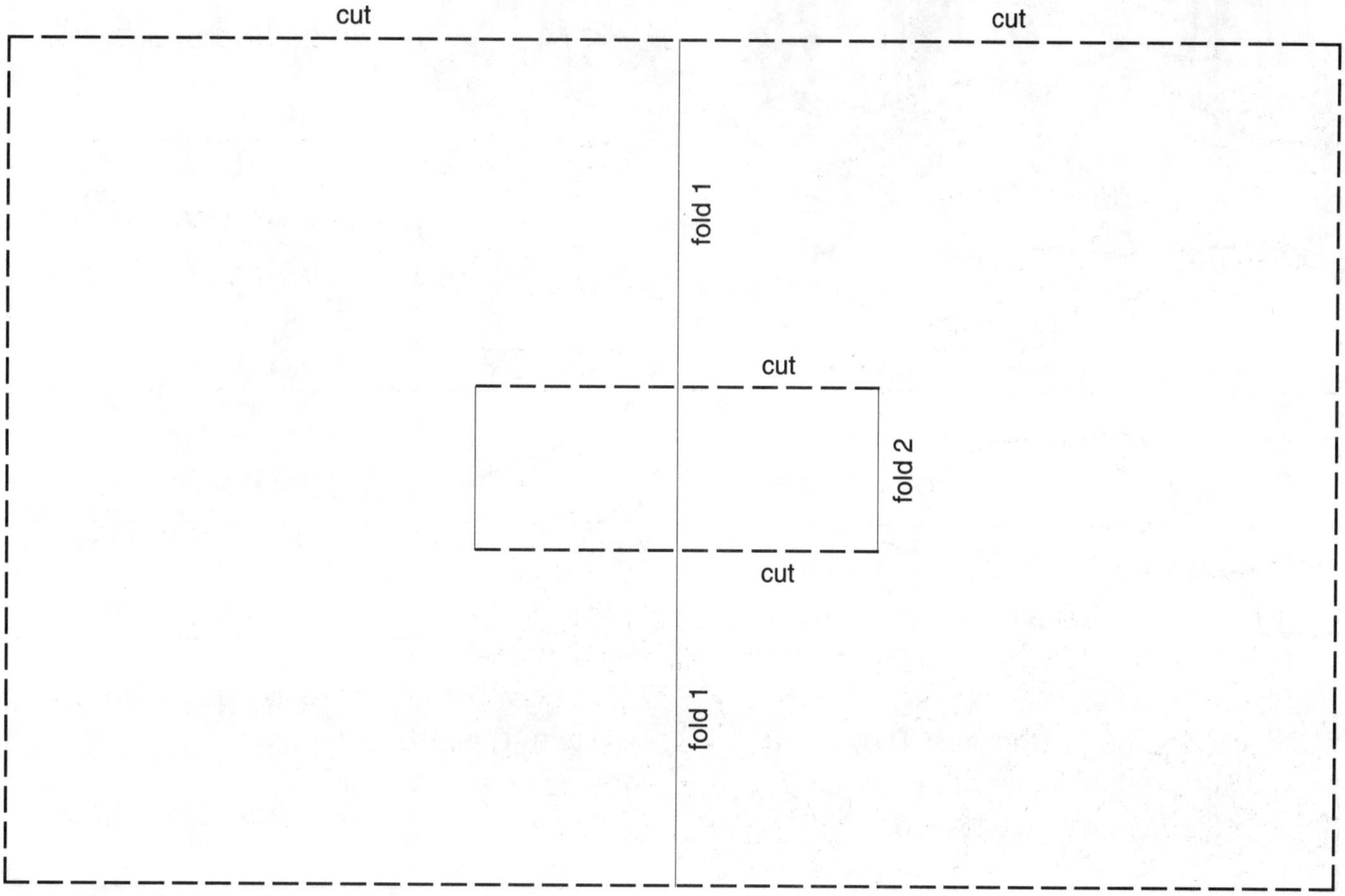

Fish

Swift Swimmers

Many different fish live in the pond.
All fish have backbones.
Fish get their oxygen from the water.

There are over 8,400 kinds of freshwater fish. Most freshwater fish are called bony fish. They have 700-800 bones in their bodies. Their excellent eyesight helps them to be good predators. They move quickly through the water by swinging their tail fins from side to side.

Fish get their oxygen from dissolved oxygen in the water. Fish gulp water in through their mouths and pump it over their gills. The gill filaments absorb the oxygen. Then the water passes out through the gill openings.

A Fish Portrait

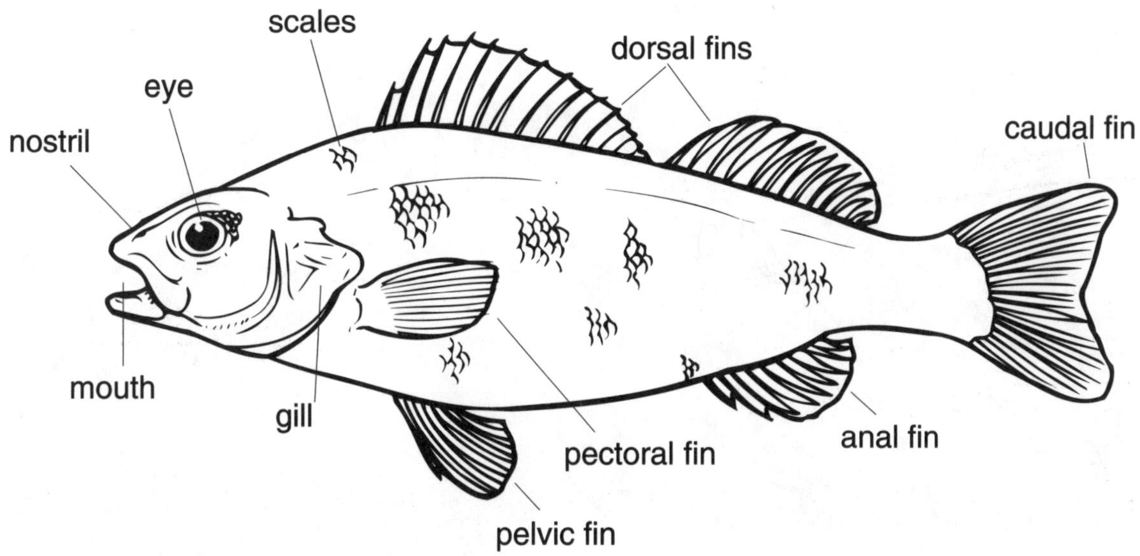

How to Make a Fish Print

Buy a small trout or bass at the grocery store or fish market. Use the fish to make prints. After students have completed the prints, encourage them to identify the important parts of the fish.

Materials:

- newsprint
- paints — tempera, acrylic, or fabric paint
- brushes
- fish
- cotton balls
- bucket with water
- newspaper
- paper towels

Steps to follow:

1. Prepare the fish for printing. Blot excess moisture from the fish with a paper towel. Stuff tiny bits of cotton under the gill and the lateral fin so they will print clearly. Place the fish on a sheet of newspaper.

2. Using a brush, paint the fish. Start with the head and work toward the tail. Make sure to cover all surfaces and edges of the fish. Lift the fish gently and place it on a clean sheet of newspaper.

3. Place the newsprint over the fish. Starting at the center of the fish, pat gently over the newsprint with your fingers.

4. Gently peel the newsprint from the fish. Hang print to dry.

5. Repaint fish for another print.

Winter in the Water

Pond animals find special ways to care for themselves during the cold winter.
Some animals migrate to warmer places.
Some animals hibernate.

During the winter, food energy is hard to find. Animals quickly use up food energy keeping warm. Pond animals adapt to the scarcity of food and the cold weather in different ways.

The fish of freshwater ponds must adjust their living habits to changes in the water temperature. If the pond freezes over during the winter, the fish move down to the warmer water near the bottom and stay there until spring.

Many ducks and other birds escape winter's hardships by migrating south. Some mammals remain active all winter long. The mink continue to hunt the shoreline. Under the ice, the muskrats emerge from their lodges and forage on plant root and stems. Beaver visit their underwater stores of stems.

For other animals, hibernation is the key to winter survival. Frogs, turtles, and salamanders burrow into the mud of the pond bottom to spend the winter. Sometimes they share the same burrow.

A Pond and Stream of Your Own

Read the Caldecott Honor Book, ***In the Small, Small Pond*** by Denise Fleming.

List the pond animals described in the book. Add other favorite pond and stream animals to the list. Paint a mural of the pond and stream, including the animals that you listed.

Materials:

- tempera paints
- paint brushes
- butcher paper
- scissors
- broad-tip black markers

Steps to follow:

1. Cover a bulletin board with butcher paper.
2. Plan the mural with the class. Discuss what should be included in addition to the animals you have already listed.
3. Decide who will paint the mural background (pond, stream, plants).
4. Assign a child to paint each animal on a piece of butcher paper (12" x 18" [30.5 x 46 cm] is a good size).
5. When animals have dried, outline with black marker and cut out.
6. Let each child decide where their animal should be placed on the mural.
7. As a class, write words describing the animals in your mural. Write the descriptions on strips of paper and staple near the illustrations

Note: Reproduce pages 44-47 for each student. Cut on dotted line and staple to create a mini-book. Students color illustrations.

A Stream Becomes A Pond

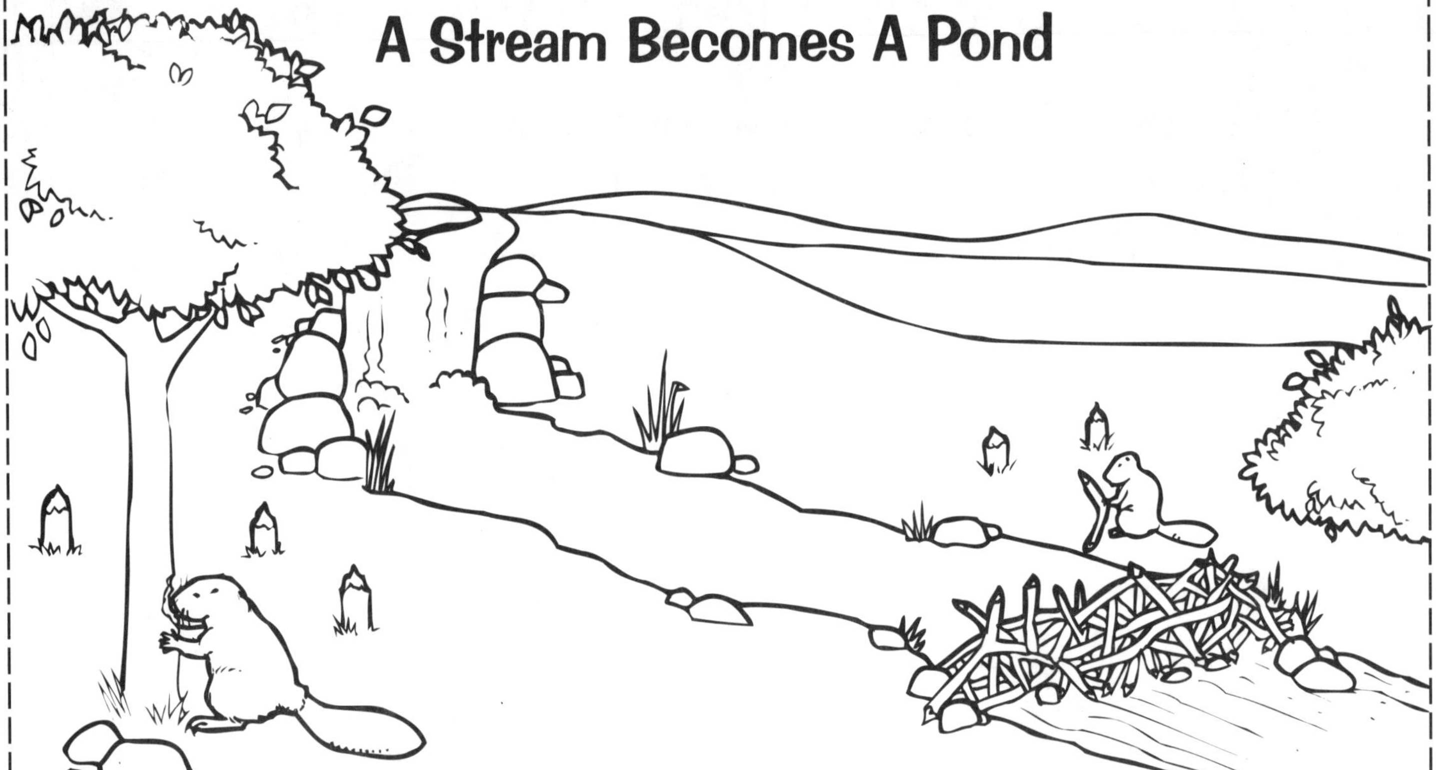

The bubbling stream rushes across the rocks.

Busy beavers build a dam.

1

A pond is formed.

By the quiet pool, plants climb, reaching for the sun.

2

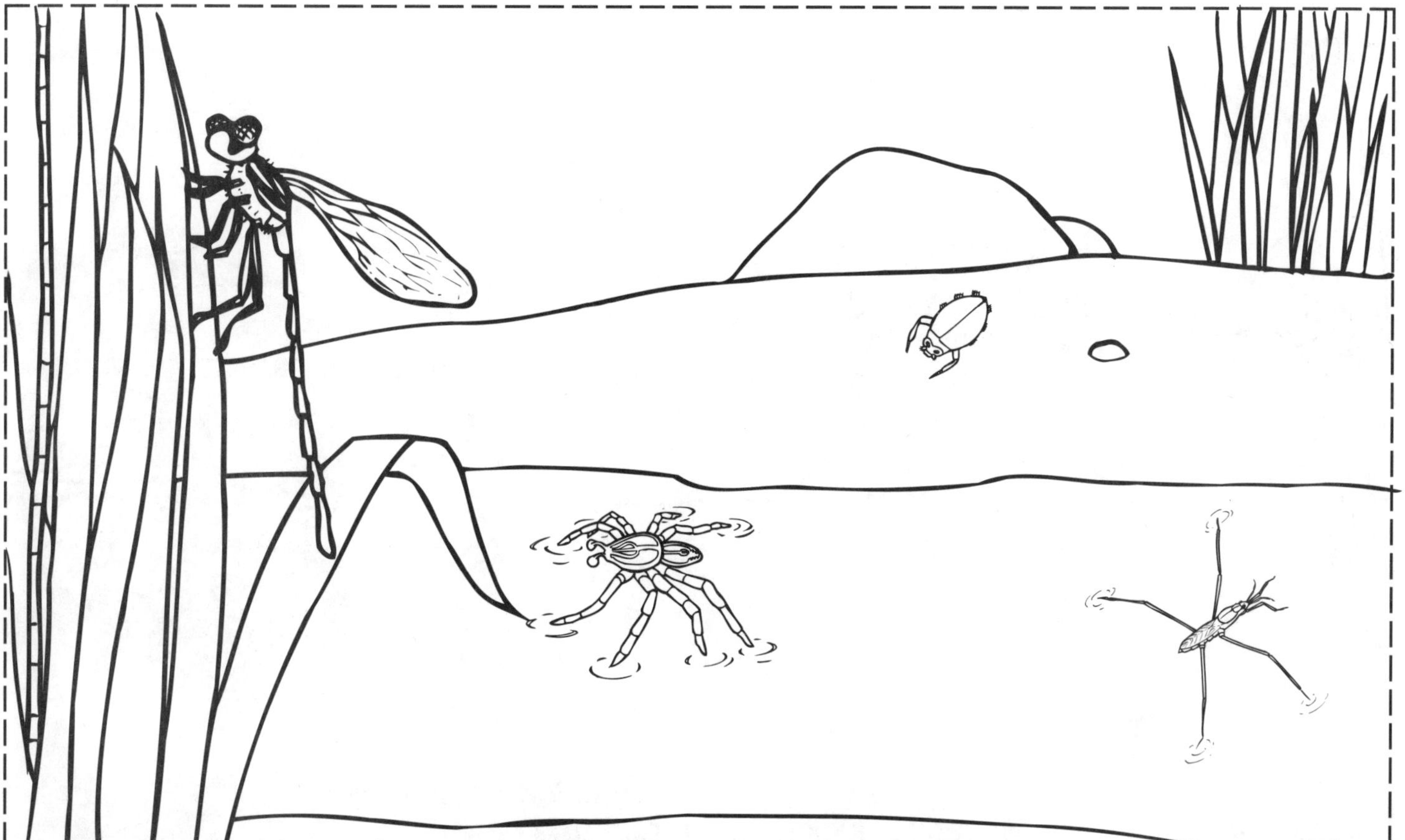

Insects buzz and hum and creep and fly.

3

Frogs and turtles hide in the reeds along the bank and swim in the shallow water.

4

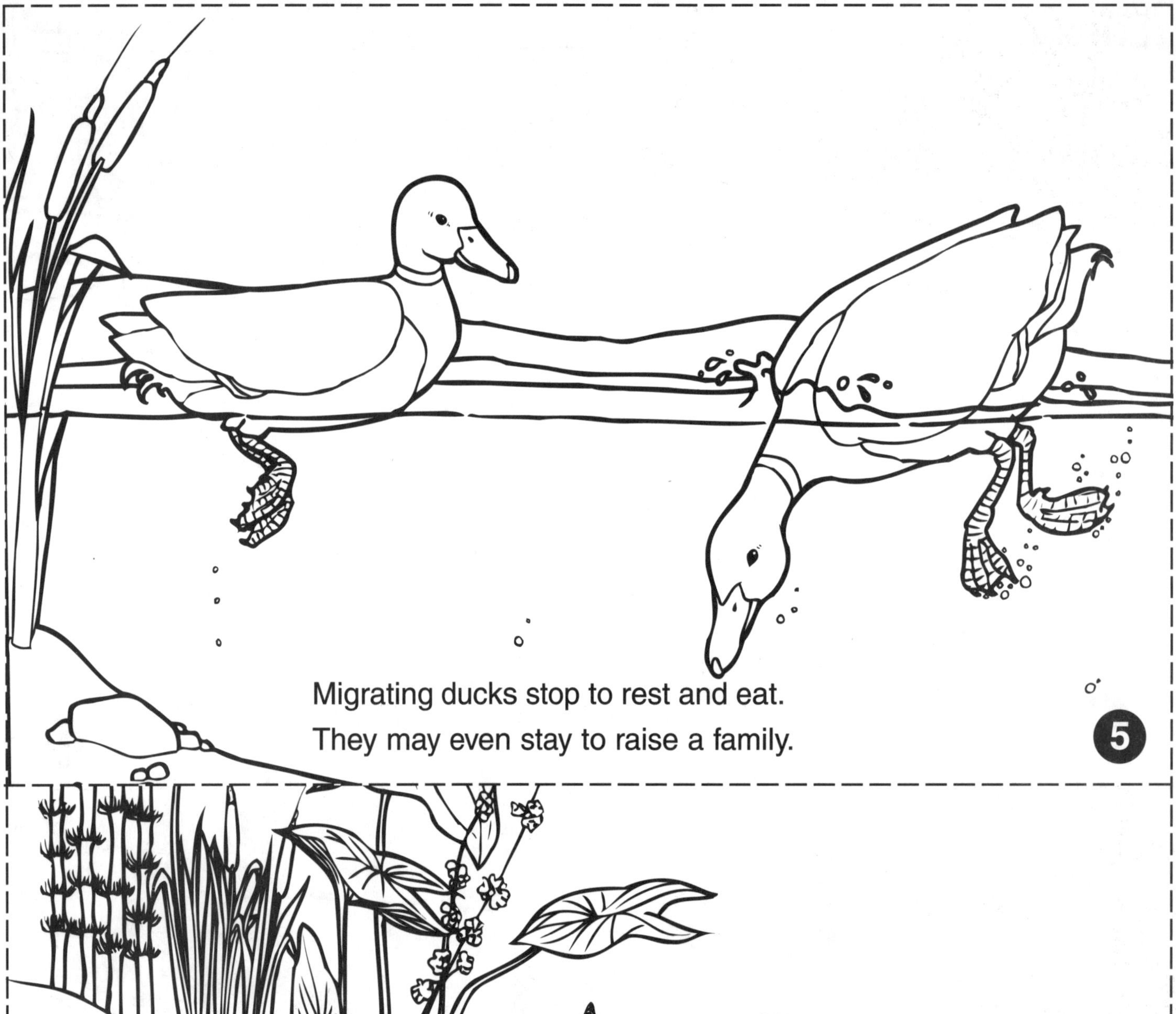

Migrating ducks stop to rest and eat.
They may even stay to raise a family.

5

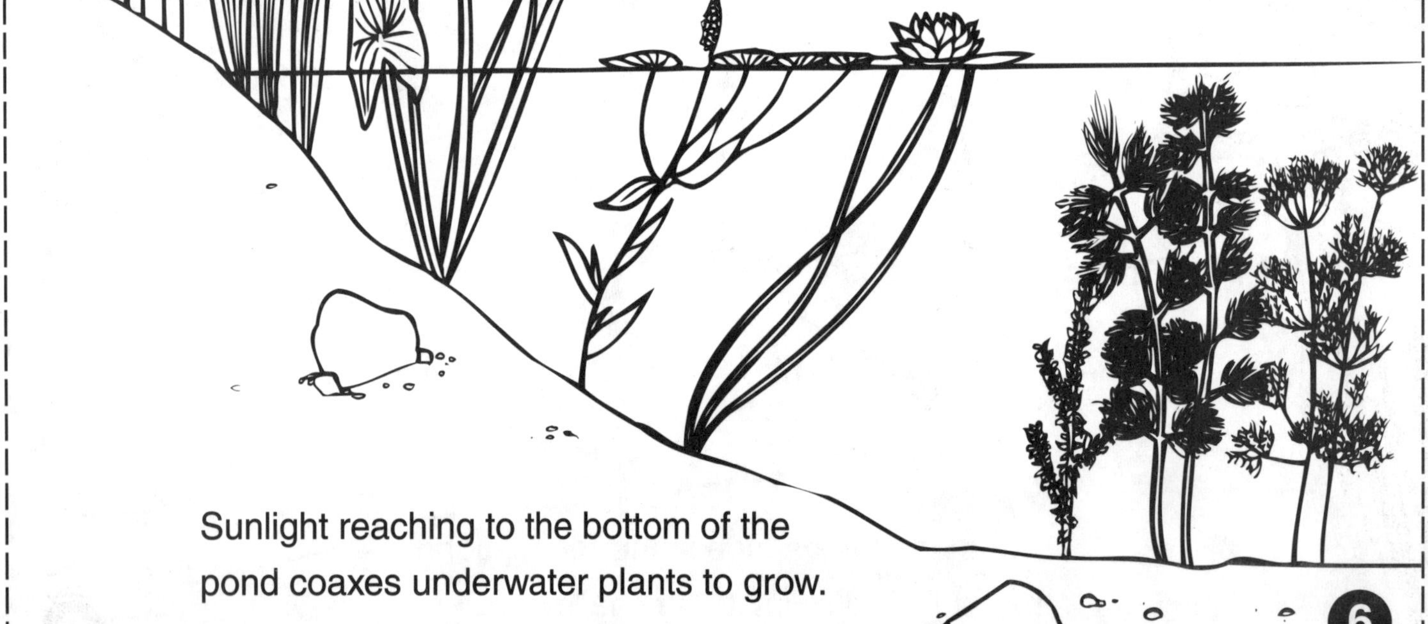

Sunlight reaching to the bottom of the pond coaxes underwater plants to grow.

6

Small fish and insects hide among the leaves.
They wait for swimming prey or hide from hungry predators.

7

Raccoons come to drink and eat,
afterwards returning to their homes.

8

Protect Pond and Stream Habitats

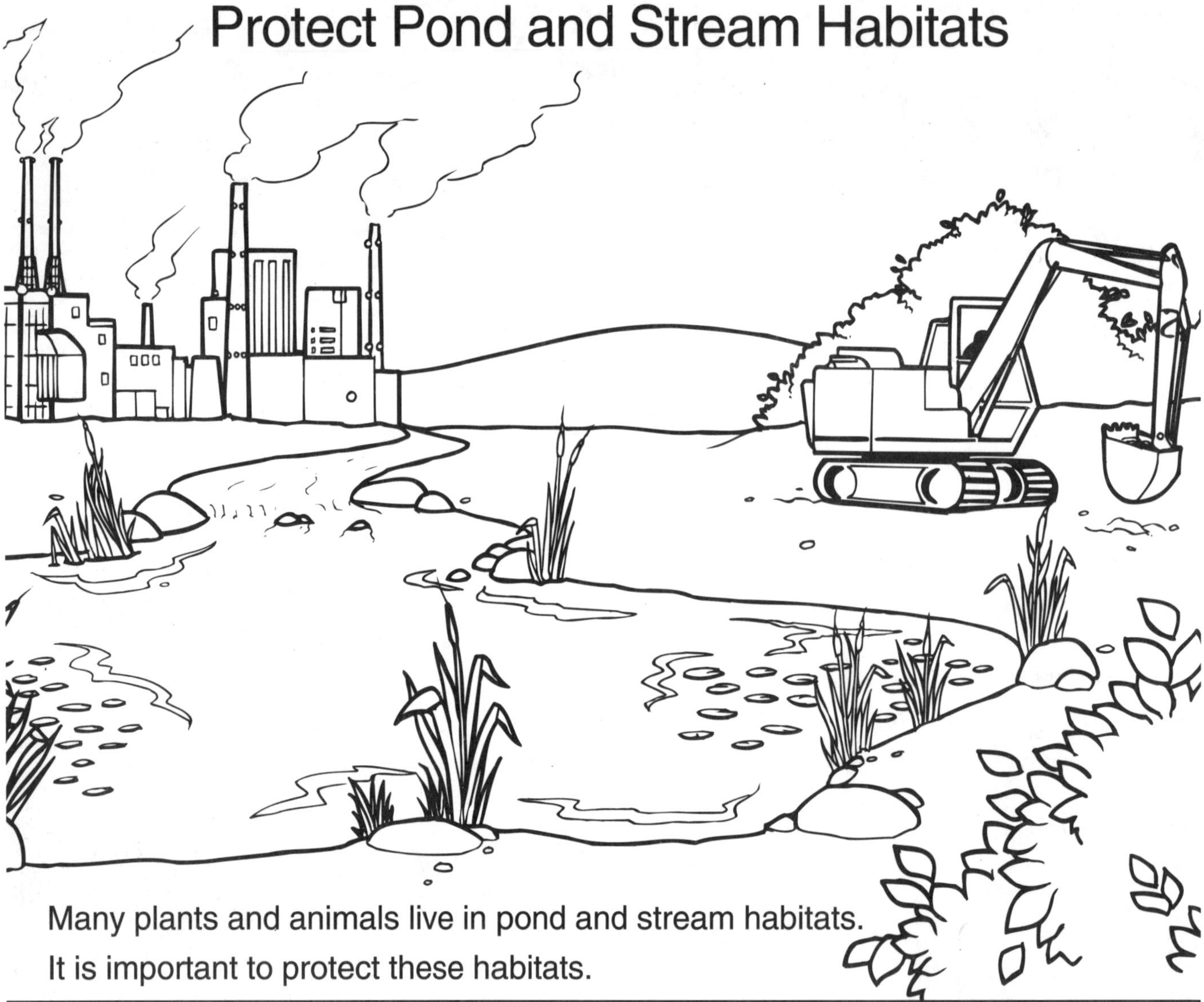

Many plants and animals live in pond and stream habitats. It is important to protect these habitats.

When people change an environment they shift the balance that exists between the living and nonliving parts of that environment. Plants and animals cannot adjust instantly to changing habitats. In the pond and stream, pollution by temperature change, fertilizer runoff, and toxic waste can change the balance.

Ponds and streams are a part of an ecological community called wetlands. Wetlands are the transition between "dry" habitats and the habitats of deep water like rivers, lakes, and oceans. Each year more wetlands are drained for agricultural purposes or filled for construction. Dams are built to create reservoirs to supply water and to generate hydroelectric power. Power plants use millions of gallons of water as a coolant and then return the heated water into rivers and streams.

1200 species of plants and over 10,000 species of animals live in pond and stream habitats. It is important to be aware of the impact of human activities on freshwater environments. The wetlands are crucial to plant and animal life. They are crucial to human life as well.